Wastewater Reuse
for
Golf Course Irrigation

Sponsored by
United States Golf Association

In Cooperation with
American Society of Golf Course Architects
Golf Course Builders Association of America
Golf Course Superintendents Association of America
National Golf Foundation

LEWIS PUBLISHERS
Boca Raton   Ann Arbor   London   Tokyo

Library of Congress Cataloging-in-Publication Data

WASTEWATER REUSE FOR GOLF COURSE IRRIGATION
Sponsored by the United States Golf Association ... [et al.].
p.cm.

Includes bibliographical references and index.

1. Sewage irrigation—Congresses.
2. Golf courses—Congresses.
I. United States Golf Association.   II. Title.

TD760.G65   1994
796.352'06'8—dc20   93-45019

ISBN 1-56670-090-6

Cover photos courtesy of James F. Moore, Director, Mid-Continent Region, USGA Green Section.

LEWIS PUBLISHERS
121 South Main Street, Chelsea, MI 48118

PRINTED IN THE UNITED STATES OF AMERICA
1 2 3 4 5 6 7 8 9 0
Printed on acid-free paper

Contents

Foreword

In 1978, four organizations in golf joined together to sponsor the first national symposium concerning the use of effluent water for golf course irrigation. The proceedings from that symposium served for many years as an important reference for golf course architects, engineers, golf course superintendents, equipment manufacturers, municipal water officials and others involved in irrigating turfgrasses with sewage facility effluent.

Fifteen years later, five golf organizations cooperated to sponsor another conference to update what has been learned about using effluent water on golf courses during the intervening years. It is clear that much has changed, both technically and politically, since the first symposium. Irrigation technology has improved significantly, and our understanding about the effects of effluent water on turfgrass systems has been greatly expanded. Politically, the use of potable water for golf course irrigation has become an important issue in many parts of the country, and the use of effluent water is mandatory in some areas.

It is also clear that there is a great need to educate regulatory officials, environmentalists, elected officials and others about the benefits and potential problems related to the use of effluent water for irrigating golf courses. It is the hope of the sponsors that this proceedings will play a significant role in the educational process.

James T. Snow
National Director
USGA Green Section

1993 Golf Course Wastewater Symposium Program Committee

USGA
James T. Snow
National Director
USGA Green Section

Michael P. Kenna, Ph. D.
Director of Research
USGA Green Section

Kimberly S. Erusha, Ph. D.
Manager, Technical Communications
USGA Green Section

ASGCA
Garrett Gill
David Rainville
American Society of Golf Course Architects

NGF
Richard C. Norton
National Golf Foundation

GCSAA
Robert Ochs
General Counsel
Golf Course Superintendents Association of America

GCBAA
Philip Arnold
Executive Vice President
Golf Course Builders Association of America

1993 Golf Course Wastewater Symposium Editorial Review Committee

James T. Snow, Chairman
National Director
USGA Green Section

Kimberly S. Erusha, Ph.D.
Manager, Technical Communications
USGA Green Section

Mike Henry
Cooperative Extension
University of California

Michael P. Kenna, Ph.D.
Director of Research
USGA Green Section

Charles H. Peacock, Ph.D.
Department of Crop and Soil Sciences
North Carolina State University

James R. Watson, Ph.D.
Vice President
The Toro Company

Preparation of List of Golf Courses Utilizing Effluent Water

American Society of Golf Course Architects

Rainbird Golf

The Toro Company

Chapter 1   Why Use Effluent?

Water—
Where Will It Come From?

James R. Watson, Ph.D.
Vice President
Toro Company

A Look at Turfgrass Water Conservation

Robert N. Carrow, Ph.D.
Professor
Crop and Soil Sciences Department
University of Georgia

Effluent for Irrigation:
The Wave of the Future?

Garrett Gill and David Rainville
Members
American Society of Golf Course Architects

1

Water—Where Will It Come From?

James R. Watson, Ph.D.
Vice President
Toro Company
Minneapolis, MN

To answer the question posed in the title we must recognize that it correctly implies concern as to availability, especially of potable water, and that there is a limited, inadequate, economic supply of all classifications of water to meet projected needs. This is true and has been recognized for some 25–30 years. Yet until severe droughts occur, the general public does not seem to be concerned and we, in the green industry, seemingly have failed in our responsibility to disseminate useful information and to utilize technologies that collectively would conserve large quantities of water.

Water—How Much? Where Does It Come From?

In many respects, water is like land. There is only so much of it and nature is not making new supplies, only recycling a portion of it. Much of the world's water supply—in excess of 99 percent—is unavailable for man's most important uses of water—drinking, manufacturing, sanitation, recreation and irrigation because it is tied up in the oceans and in the polar ice caps[5].

The amount of water on earth is essentially non–destructible and fixed.

The amount of water on earth is essentially non-destructible and fixed. Certain types of volcanic activities add small amounts of new water, but it is insignificant in the overall scheme (Table 1, Figure 1).

Worldwide, some 80,000 to 85,000 cubic miles of water evaporate each year from oceans and another 15,000 to 17,000 cubic miles from lakes, streams, and land surfaces[2]. Some 24,000 cubic miles of this atmospheric

Table 1. World Water Balance: Estimated World Water Supply and Budget

Water Items	Volume (Thousands) Cubic Miles	Cubic Kilometers	Percent Total Water
Water in land areas:			
Fresh-water lakes	30	125	0.009
Saline lakes and inland seas	25	104	0.008
Rivers (average instantaneous volume)	0.3	1.25	0.0001
Soil moisture and vadose water	16	67	0.005
Ground water to depth of 4,000 m (about 13,100 ft.)	2,000	8,350	0.61
Icecap and glaciers	7,000	29,200	2.14
Total in land area (rounded)	9,100	37,800	2.8
Atmosphere	3.1	13	0.001
World ocean	317,000	1,320,000	97.3
Total, all items (rounded)	326,000	1,360,000	100
Annual evaporation[1]:			
From world ocean	85	350	0.025
From land areas	17	70	0.005
Total	102	420	0.031
Annual precipitation:			
On world ocean	78	320	0.024
On land areas	24	100	0.007
Total	102	420	0.031
Annual runoff to oceans from rivers and icecaps	9	38	0.003
Ground-water outflow to oceans[2]	0.4	1.6	0.0001
Total	9.4	39.6	0.0031

[1]Evaporation (420,000 Km³) is a measure of total water participating annually in the hydrologic cycle
[2]Arbitrarily set equal to about 5 percent of surface runoff.

Source: Nace, U.S. Geological Survey, 1967

Figure 1. Water Availability on Earth Source: Doxiadis, C.A., 1967, Water and Environment, International Conference on Water for Peace, Washington, DC

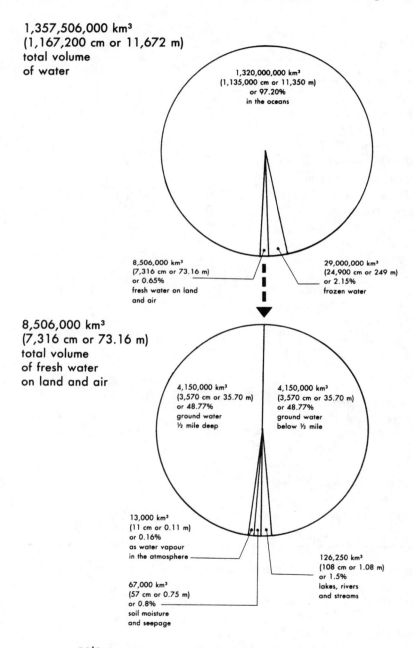

1,357,506,000 km³
(1,167,200 cm or 11,672 m)
total volume
of water

1,320,000,000 km³
(1,135,000 cm or 11,350 m)
or 97.20%
in the oceans

8,506,000 km³
(7,316 cm or 73.16 m)
or 0.65%
fresh water on land
and air

29,000,000 km³
(24,900 cm or 249 m)
or 2.15%
frozen water

8,506,000 km³
(7,316 cm or 73.16 m)
total volume
of fresh water
on land and air

4,150,000 km³
(3,570 cm or 35.70 m)
or 48.77%
ground water
½ mile deep

4,150,000 km³
(3,570 cm or 35.70 m)
or 48.77%
ground water
below ½ mile

13,000 km³
(11 cm or 0.11 m)
or 0.16%
as water vapour
in the atmosphere

67,000 km³
(57 cm or 0.75 m)
or 0.8%
soil moisture
and seepage

126,250 km³
(108 cm or 1.08 m)
or 1.5%
lakes, rivers
and streams

note: figures in brackets indicate the height that the relevant quantities of water would reach if they were placed on the whole non-frozen land area of the earth which is 116,400,000 km²

water falls on land surfaces in the form of precipitation. This is enough water to form a global lake 33 feet deep[5].

It is further estimated that, on average, between 600 and 675 billion gallons of fresh water falls on the contiguous 48 states daily[3,7]. This equates roughly to 30 inches of annual rainfall.

Hydrologic Cycle

This solar fired, water cycling process is a continuous, never-ending cycle of charging (evapotranspiration) and discharging (precipitation). It is this process that has caused the same water to be used and reused since time on earth began (Figure 2).

The amount of evaporation from water and land sources and the amount transpired by plants is approximately equal to the amount that falls daily as precipitation, i.e. rain, snow, sleet, hail, fog, dew. Precipitation patterns, however, are widely variable and seldom, if ever, is the right amount of water, in any form, delivered to the place where it is needed or not needed, as the case may be.

Human activity, while often lowering water quality and frequently displacing water from natural channels, has little effect on the hydrologic cycle (Figure 3). In the future, through weather modification, humans may disrupt the hydrologic cycle or affect precipitation in a specific location. Laws governing the atmospheric water segment of the hydrologic cycle continue to evolve[4].

Source and Fate of Precipitation

Precipitation that falls on land surfaces infiltrates the earth's surface to provide water for plants and other soil borne organisms, replenishes reservoirs of underground water (aquifers) or runs off into lakes and streams where some of it evaporates. Some of this precipitation eventually returns to the oceans.

Of the 40,000 billion gallons daily (bgd) of water vapor that passes over the conterminous 48 states, approxi-

Human activity, while often lowering water quality and frequently displacing water from natural channels, has little effect on the hydrologic cycle.

Precipitation that falls on land provides water for
- *plants and soil borne organisms*
- *underground reservoirs (aquifers)*
- *lakes and streams*

5

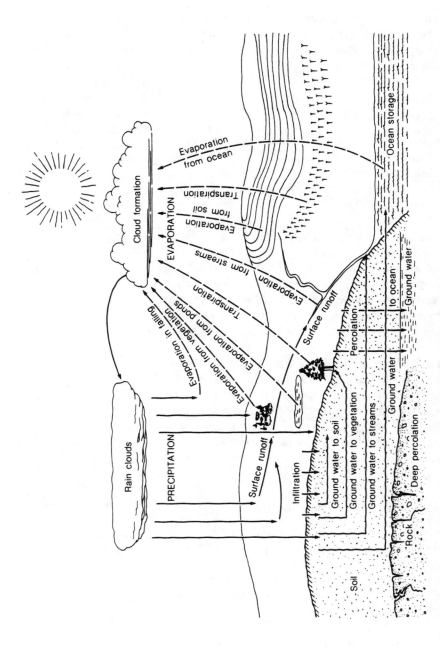

Figure 2. The Hydrologic Cycle: A Descriptive Representation Source: Council on Environmental Quality, 1981

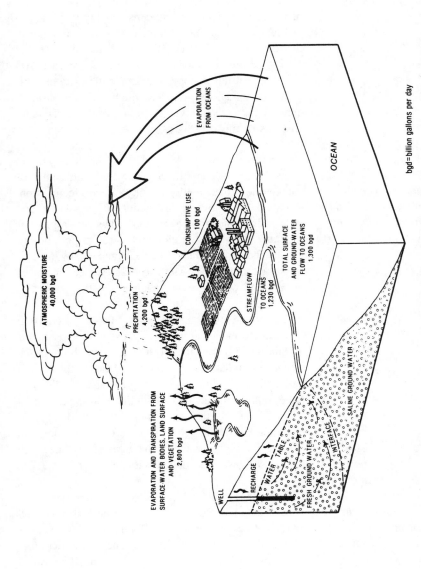

Figure 3. Hydrologic Cycle Showing the Gross Water Budget of the Conterminous United States

Source: U.S. Geological Survey, 1984, National Water Summary 1983, Hydrologic Events and Issues, Water-Supply Paper 2250

bgd=billion gallons per day

ATMOSPHERIC MOISTURE
40,000 bgd

EVAPORATION
FROM OCEANS

OCEAN

PRECIPITATION
4,200 bgd

EVAPORATION AND TRANSPIRATION FROM
SURFACE WATER BODIES, LAND SURFACE
AND VEGETATION
2,800 bgd

CONSUMPTIVE USE
100 bgd

STREAMFLOW
TO OCEANS
1,230 bgd

TOTAL SURFACE
AND GROUND-WATER
FLOW TO OCEANS
1,300 bgd

WELL

RECHARGE

WATER TABLE

FRESH GROUND WATER

INTERFACE

SALINE GROUND WATER

Insufficient storage areas and the extreme variability of annual precipitation limits recapture of greater amounts of run–off waters.

mately 10 percent (4,200 bgd or approximately 30 inches annually) falls as precipitation (Figure 4). Two thirds of this amount (2,750 bgd) evaporates immediately, or is transpired. The remainder accumulates in surface or ground storage areas, is used consumptively, evaporates or runs off into the Atlantic and Pacific oceans, the Gulf of Mexico, or rivers and streams flowing into Canada or Mexico. Table 2 shows the distribution of water in the coterminous 48 states. Only 675 bgd of the potential 1,450 bgd are considered available in 95 of each 100 years[6]. Insufficient storage areas and the extreme variability of annual precipitation which results in drought or floods currently limits utilization or recapture of greater amounts of run–off waters.

Urbanization plays a major role in recharge of ground water and excess run–off.

Urbanization plays a major role in recharge of ground water and excess run–off. Sites covered with asphalt or concrete lose their ability to fully utilize precipitation. Removal of trees and destruction of grasslands and farmlands, along with paving these sites with asphalt or concrete, disrupts infiltration and increases pollution of ground and surface waters. Table 2, prepared by the U.S. Geological Survey, shows a selected sequence of changes in land and water use associated with urbanization.

Precipitation patterns in the lower 48 states range from less than 10 inches in the West and Southwest to over 100 inches in the Pacific Northwest. Two thirds of the 30 inches annual average rainfall occurs in the eastern half of the U.S. (Figure 4). The approximate dividing line is the 100th meridian. West of this line, with the exception of the Pacific Northwest, evaporation exceeds precipitation; east of the line precipitation exceeds evaporation (Figure 5). Figure 6 shows the areas of the U.S. with natural surpluses and deficiencies of water.

If the western half of the U.S. is to continue to grow, develop and expand as it has in the past, it must find additional sources of water.

If the western half of the U.S. is to continue to grow, develop and expand as it has in the past, it must find additional sources of water. Seventeen western states contain 85 percent of the land irrigated for agricultural purposes. Too, a large portion of the Sunbelt is located in the water-short Southwest, an area of rapidly increasing population. Where will the needed water come from? Certainly the geopolitical implications of unequal rainfall

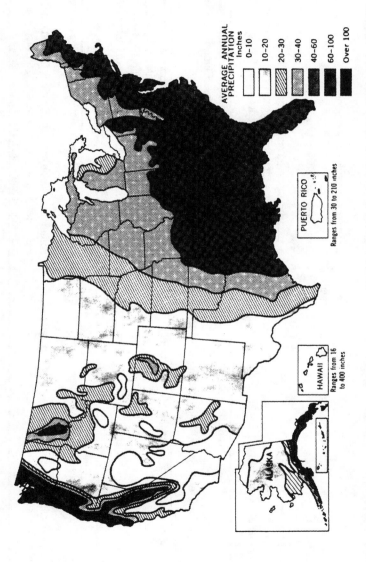

Figure 4. Average Annual Precipitation for the United States Source: U.S. Water Resources Council

Average annual rainfall values for any given area often are misleading. Wide variability in rainfall occurs seasonally, annually and, in some instances in areas of close proximity. An example of the later occurs in Kawai, Hawaii; Mount Waialeale averages more than 400 inches of annual rainfall; 25 miles to the southwest on the leeward side of the island, rainfall is less than 20 inches.

Table 2. Hydrologic Effects of Urbanization: A Selected Sequence of Changes
in Land and Water Use Associated with Urbanization Source: U.S. Geological Survey

Changes in Land or Water Use	Possible Hydrologic Effect
Transition from pre-urban to early-urban stage: Removal of trees or vegetation	Decrease in transpiration and increase in storm flow.
Construction of scattered city-type houses and limited water and sewage facilities.	Increased sedimentation of streams.
Drilling of wells	Some lowering of water table.
Construction of septic tanks and sanitary drains	Some increase in soil moisture and perhaps a rise in water table. Perhaps some waterlogging of land and contamination of nearby wells or streams from over-loaded sanitary drain system.
Transition from early-urban to middle-urban stage: Bulldozing of land for mass housing, some topsoil removed, farm ponds filled in.	Accelerated land erosion and stream sedimentation and aggradation. Increased flood flows. Elimination of smallest streams.
Mass construction of houses, paving of streets, building of culverts	Decreased infiltration, resulting in increased flood flows and lowered groundwater levels. Occasional flooding at channel constrictions (culverts) on remaining small streams. Occasional overtopping or undermining of banks of artificial channels on small streams.
Discontinued use and abandonment of some shallow wells	Rise in water table.
Diversion of nearby streams for public water supply.	Decrease in runoff between points of diversion and disposal.
Untreated or inadequately treated sewage discharged into streams or disposal wells.	Pollution of stream or wells. Death of fish and other aquatic life. Inferior quality of water available for supply and recreation at downstream populated areas.

Table 2. Hydrologic Effects of Urbanization: A Selected Sequence of Changes in Land and Water Use Associated with Urbanization (continued)

Transition from middle-urban to late-urban stage:	Reduced infiltration and lowered water table. Streets and gutters act as storm drains, creating higher flood peaks and lower base flow of local streams.
Urbanization of area completed by addition of more houses and streets and of public, commercial, and industrial buildings.	
Larger quantities of untreated waste discharged into local streams	Increased pollution of streams and concurrent increased loss of aquatic life. Additional degradation of water available to downstream users.
Abandonment of remaining shallow wells because of pollution	Rise in water table
Increase in population requires establishment of new water supply and distribution systems, construction of distant reservoirs to divert water from upstream sources within/outside basin	Increase in local stream flow if supply is from outside basin.
Channels of streams restricted at least in part to artificial channels and tunnels	Increased flood damage (higher stage for a given flow).
	Changes in channel geometry and sediment load.
	Aggradation.
Construction of sanitary drainage system and treatment plant for sewage	Removal of additional water from the area, further reducing infiltration and recharge of aquifer.
Improvement of storm drainage system	A definite effect is alleviation or elimination of flooding of basements, streets, and yards, with consequent reduction in damages, particularly with respect to frequency of flooding.
Drilling of deeper, large-capacity industrial wells	Lowered water-pressure surface of artesian aquifer; perhaps some local overdrafts (withdrawal from storage) and land subsidence. Overdraft of aquifer may result in salt-water encroachment in coastal areas and in pollution or contamination by inferior or brackish waters.

Table 2. Hydrologic Effects of Urbanization: A Selected Sequence of Changes in Land and Water Use Associated with Urbanization (continued)

Increased use of water for air conditioning	Overloading of sewers and other drainage facilities.
	Possibly some recharge to water table, due to leakage of disposal lines.
Drilling of recharge wells	Raising of water-pressure surface
Wastewater reclamation and utilization	Recharge to groundwater aquifers.
	More efficient use of water resources.

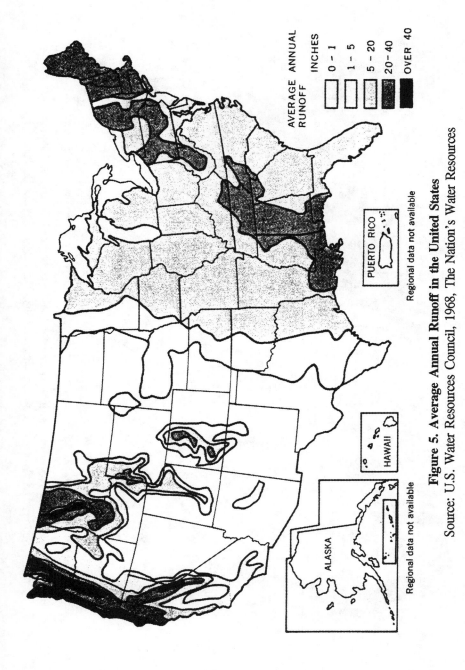

Figure 5. Average Annual Runoff in the United States
Source: U.S. Water Resources Council, 1968, The Nation's Water Resources

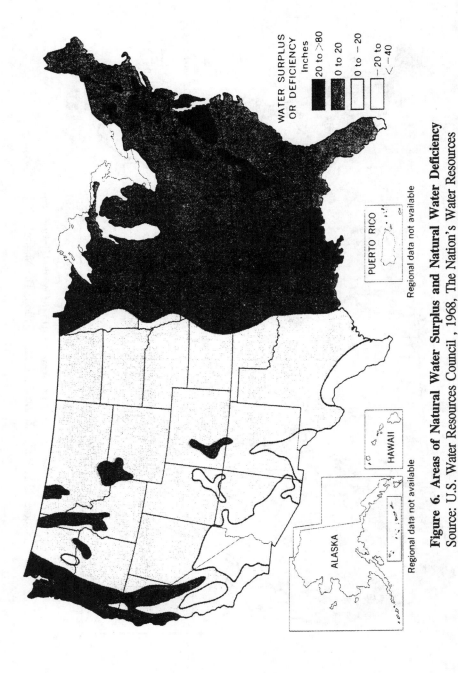

Figure 6. Areas of Natural Water Surplus and Natural Water Deficiency
Source: U.S. Water Resources Council , 1968, The Nation's Water Resources

distribution and allocation of surface and ground waters are enormous! And although schemes abound for re-distribution from the east and north to the west and southwest, none at present are plausible or economical.

Usage

Agriculture uses in excess of 80 percent of the fresh water consumed in the U.S. As populations increase and demands for more food and fiber increase, we can expect a corresponding increase in the demand for fresh water for all purposes. Unfortunately, unless we plan properly, seek alternative sources, develop marginal waters and learn how to use them and develop and use plants that tolerate saline and sodic waters, this may come at the expense of the water needed to support and sustain green spaces. Green spaces provide for healthful recreational pursuits, control dust and erosion, help to remove air pollutants and help conserve energy through the cooling effects of evaporation in the summer and serving as wind breaks in the winter.

Agriculture uses in excess of 80 percent of the fresh water consumed in the U.S.

Where Will the Water Come From?

So, what can be done? Where will the water come from? How do we "get a handle" on this most essential and basic resource? The answer, I believe, lies in six general areas.

First, expand efforts to develop and enhance storage areas capable of capturing greater volumes of flood and storm waters. These waters currently are wasted for the most part. They do, however, represent a substantial source of water that currently is not used.

Second is conservation, which embodies not only the standard techniques and procedures that conserve water, but also encompasses the area concerned with proper management of water. This category is sometimes separated from conservation, but from the standpoint of this discussion, I have chosen to group the two.

What can be done?
- *expand storage and capture areas*
- *improve conservation*
- *research to produce better plants*
- *utilize wastewater*
- *desalinize sea water*
- *use agricultural water more efficiently*

Third, water will be saved through research, especially that concerned with development of new plants—both economic crops and turfgrasses. We must develop plants that will function better under stress and tolerate marginal resources such as poor soils or brackish water.

Fourth, we must begin to utilize wastewater—water that today is, in most cases, a resource simply wasted. This is the basic theme of this symposium.

Fifth, desalinization of seawater when economically feasible will supply water for cities and towns adjacent to oceans and seas. Some towns already use this source.

Sixth, water currently allocated to agriculture will be partly reallocated to other essential uses.

Water Conservation

We must save, or reuse, or recycle the water available to us. When applied to turfgrass management, it boils down to some very fundamental, down–to–earth changes in habits and practices.

Every turfgrass manager in America, in the entire world for that matter, must begin immediately to plan and implement a program of water conservation and wise management of this finite resource.

Every turfgrass manager must begin immediately to plan, and implement a program of water conservation and wise management of this finite resource.

The immediate result will be a savings in costs. Less water will be consumed which will result in a reduction in the dollar amount spent for the water and the cost to pump and transport it. But more important, far more important, it will help the turf manager prepare for the day when it will become necessary for him to function with less water, either because the water simply will not be there, physically, in the current quantities or because someone . . . some bureaucratic or legislative authority . . . will have diverted it to a purpose considered more politically important.

There are six simple, basic steps that will lead to water conservation and improved management. They often receive entirely too little attention and respect.

1. Establish watering priorities. Give highest priority to the most intensively managed areas; for example, the putting greens, the most valuable part of the golf course and where the most critical play takes place.

2. Follow sound irrigation practices. Irrigate when there is the best combination of reduced wind, low temperature and high humidity. When watering trees and shrubs, use soil probes so the water will penetrate deeply. Preferably, install drip systems for these types of plants.

3. Reduce, or avoid where possible, other causes of stress. Make certain there is adequate internal soil drainage to ensure maximum root growth and, more importantly, avoid rootzone saturation.

4. Alter major cultural practices. (1) Test the soil annually to ensure adequate fertility, especially for phosphorus, which encourages deeper roots, thus expanding the area from which the turfgrass can draw nutrients and moisture. (2) Raise the height of cut for all areas. Raising the height of cut on a golf course green as little as 1/32 of an inch can have significant effect on the ability of the grass to tolerate stress and make more efficient use of soil moisture. (3) Increase the frequency of spiking or cultivation to trap moisture and hold it longer in the vicinity of the root system.

5. Expand the use of mulch. This is often overlooked. Apply heavy layers of mulch (compost, grass clippings, leaves) or any organic debris that is available around the base of trees, shrubs and flower beds, to reduce evaporation and to help control weeds.

6. Erect wind barriers, especially where there are large expanses of open spaces.

Six steps to water conservation:
- *prioritize water use*
- *follow sound irrigation practices*
- *reduce plant stress*
- *alter major cultural practices*
- *expand the use of mulches*
- *erect wind barriers*

Water Management

One of the major challenges facing the green industry is to find a way to narrow the gap between what we know about water management and what we practice.

Many involved in the turf management business are guilty of misusing water. That's true, but to a far lesser degree on golf courses than for most of the other major turf facilities. Yet, golf courses do not take full advantage of many of the technical accomplishments of the irrigation industry. For example, each year we lose grass, sometimes a great deal of grass, due to the mismanagement of water. Irrigation management and irrigation equipment, i.e. sprinkler heads, valves, controllers and pumps and their controls, are capable of saving large quantities of water by controlling its application based on scientific information. Both areas contribute to water conservation and are a potential source of water.

One of the major challenges facing the green industry, and golf courses in particular in the next few years, is to find a way to narrow the gap between what we know and what we practice. We simply must find a solution to this problem. Although it is recognized that we need to learn more about such things as drought tolerance and rooting characteristics of grasses, water requirements, watering techniques, water conservation, soil-air-water relationships, leaching, and weeds and their ecological relationship in the turfgrass environment; we also must learn to apply the available scientific and practical information. We must move from the art of water management to the science of water management. The consequence of not applying what we know in these areas is that we are borrowing a major source of water from the future and missing an opportunity to become leaders as water conservation authorities.

We must move from the art of water management to the science of water management.

In addition to recommending that a solution to the information gap be found, there is a need to do everything possible to generate more knowledge—more new information, better technology, better products, better equipment—so that turfgrass management will continue to advance. This leads to the third area or third potential source for the water that will be needed to meet our projected needs.

Research

Fundamental, basic research to expand our knowledge on all aspects of water and its use must be supported. Further, that knowledge must be disseminated in a timely and efficient manner.

Research in related areas that will result in lower water use rates must be encouraged. For example, grasses with lower water requirements, with deep root systems and grasses that remain green and playable under drought conditions will meet this requirement. We must have salt and alkali tolerant grasses that are able to survive adverse soil conditions and that flourish when irrigated with marginal brackish, power plant blowdown, or saline waters. We must move away from the quick approach, or the applied approach to turfgrass research. It is time to step back and develop a basic foundation on the principles of turf water use from which to move or advance our industry these next several decades. This basic approach must include, among others, bio–engineering, development of transgenic grasses, precise modification of rootzones, and cultural practices that ensure quality playing surfaces.

Research is needed to develop grasses with
- *lower water requirements*
- *deeper root systems*
- *ability to remain green during drought*
- *tolerance to brackish waters*
- *tolerance to adverse soil conditions*

This approach will prove to be a valuable indirect source of water and will help to answer the question: "from where will it come?"

Wastewater

The fourth area to which we must look for tomorrow's water is the so–called wastewater, or recycled, multiple use water; water that may not be potable, but which is perfectly suitable for growing plants. Major droughts provide an impetus for use of wastewaters. People who are recycling effluent for irrigation purposes are no longer reticent about using this source.

Golf courses are by far the leading recreational turfgrass user of wastewater. The reasons are simple: golf courses have intensively managed turf, they are managed with the highest level of professionalism, and they require significant amounts of water. Another reason for in-

Golf courses are by far the leading recreational turfgrass user of wastewater.

creased use of effluent on turfgrass areas is the fact that the grass helps to utilize nutrients in the effluent. Also, most turf areas have relatively permeable soils. Usually 3 to 4 feet of soil is adequate to cleanse, purify or reclaim wastewaters. Unfortunately, agencies and individuals disagree as to the depth of soil, or distance above groundwater, that must exist before effluent may be discharged. For example, EPA says at least 5 feet of soil is needed, while the Corps of Engineers suggests 5 to 10 feet, and the surgeon general recommends 7 feet. The state of Minnesota requires 10 feet, as does Pennsylvania. Standards are needed to fully exploit this potential source.

Desalinization

Fifth, desalinated water will be obtained from the seas, oceans and saline or sodic underground waters. This area represents a major source of fresh water, especially for those adjacent to potential sources. For the most part, desalinization is not economically practical today. However, the technology is available and is being used in some cities.

Distillation, evaporation, and reverse osmosis are major techniques that will produce fresh water from salty water.

Distillation, evaporation, and reverse osmosis are major techniques that will produce fresh water from salty water. Distillation requires energy from fossil fuels or from burning of wood. Evaporation normally employs sunlight. The major draw back is the large areas required to capture adequate energy from the sun to effect evaporation on a scale large enough to be practical.

Reverse osmosis is practical, but somewhat expensive. This process is used in the Middle East and other regions throughout the world to filter saline water. In some cases where the water is used to irrigate golf courses, reverse osmosis reduces the salt content from 13 to 14,000 ppm to 6 to 7000 ppm. Sometimes this water is diluted with "fresh" water for irrigating parks and other landscaped areas.

Agricultural Water

Sixth, water currently used by agriculture will be real-located. It seems inevitable that the 80 to 85 percent of fresh water currently allocated for agriculture will be reduced and made available for other purposes. As populations grow in states with water shortages, and as more and more fruit and vegetables are imported, agri-cultural water will cost more and more. And, this water will be diverted for purposes, other than agriculture, which demand potable water.

It seems inevitable that the 80 to 85 percent of fresh water cur-rently allocated for agriculture will be reduced and made available for other purposes.

Blending of fresh water with saline and sodic waters offers an opportunity to conserve fresh and potable water supplies. Dilution of partially desalinated water, as well as some effluent water with fresh water, provides a means of extending and conserving fresh water. Diluted nonpotable waters often may be used on golf courses, parks, and other landscaped areas. When salt tolerant plants become available, substantial quantities of fresh water may be saved through blending.

Summary

Water—where will it come from? I've listed six potential areas: (1) storage and utilization of flood and storm waters, (2) conservation including learning to apply the scientific and technical agronomic water management knowledge available to us, (3) basic research, (4) utilization of wastewater, recycled water—the subject of this sympo-sium, (5) desalinized water,and (6) water from that currently used by agriculture.

The western United States receives only one-third of the nation's average annual rainfall, yet uses enormous quantities of water, 80 to 85 percent of the nation's fresh water for irrigation— primarily for agricul-ture.

The United States has an abundant, currently adequate, and generally dependable supply of fresh water from surface and groundwater sources. On the average, in excess of 600 billion gallons per day of fresh water are available for beneficial uses.

Major problems are associated with the unequal distri-bution of precipitation, with its attendant problems of location and availability to population centers on a

Where will it come from?
- *better storage and utilization of rainfall*
- *conservation*
- *basic research*
- *wastewater*
- *desalinized water*
- *water from agriculture*
- *blending potable and nonpotable water*

Conservation and recycling of water, along with improved management and technology will contribute to more efficient use.

Turfgrass interests, especially golf courses, are turning increasingly to irrigation with wastewaters.

nationwide basis. The western United States receives only one-third of the nation's average annual rainfall, yet uses enormous quantities of water, 80 to 85 percent of the nation's fresh water for irrigation—primarily for agriculture. The West has abundant energy reserves in the form of shale oil and coal. Vast quantities of water are needed to process these and, perhaps, transport them to the energy–poor East and Midwest. Specific examples of the problem of location, other than those discussed, are given by Wiegner[7]. She points out that in Louisville, Kentucky, the groundwater table rose 32 feet between 1974 and 1979; at the same time in parts of Arizona the water table dropped 400 to 450 feet.

This inequitable distribution of available fresh water supplies at points far distant from where they are needed, or where they may be used beneficially for large numbers of people, will inevitably give rise to complex social, economic and legal problems. Certainly, water will cost much more in the future. And in all likelihood, it will be transported or otherwise relocated far distances to support human and domestic needs.

Conservation and recycling of water, along with improved management and technology, will contribute to more efficient use. Research that leads to a better understanding of our water resources and how to manage them, coupled with development and utilization of drought and salinity–tolerant plants, will contribute further to wise use of our water.

Finally, programs and facilities that will lead to widespread use of both potable and nonpotable waters will conserve water and reduce its cost. Turfgrass interests, especially golf courses, are turning increasingly to irrigation with wastewaters. Others within the green industry must follow suit. Development of dual–plumbing systems to accommodate both classes of water and to blend them for turf facilities, including home lawns, is inevitable. Most golf courses already use such systems.

Economic, social, and environmental constraints arising from water demands need not be incompatible with each other. These interrelationships are complex and will be

22

difficult to resolve. And, as the pressures on water allocation resulting from population growth and industrial or agricultural expansion become more intense, the need to conserve and wisely use the nation's bountiful supplies of fresh water cries out for reasonable resolution.

The nation's turfgrass facilities are a national treasure and must not be overlooked in the development and execution of water policy.

The nation's turfgrass facilities are a national treasure and must not be overlooked in the development and execution of water policy. They contribute to the economy and recreational value of our communities and they are aesthetically appealing and functionally beneficial.

References

1. Ackerman, W.C., E.A. Colman, and Harold Ogrosky. 1955. Where we get our water: From ocean to sky to land. Yearbook of Agriculture. USDA, Washington, DC. p. 41-51.

2. Carhart, A. H. 1951. Water or your life. J.B. Lippincott Co., New York, NY.

3. Cook, G. 1980. The thirsty high plains. Journal of Freshwater, Vol. 4. Special Issue: A decade for decisions. The Freshwater Foundation, Navarre, MN. p. 513-517.

4. Thomas, W. A. 1981. Allocation of water as a scarce resource. Futurics. 5(1):31-37 Special Issue: Water, human values and the 80s. A consortium effort to keep our waters usable. Pergamon Press, New York, NY.

5. Thompson, K. 1982. Food for the table: will water set the price? Journal of Freshwater. Vol. 6 Special Report. The Freshwater Foundation, Navarre, MN.

6. U.S. Water Resources Council. 1978. The nation's water resources, 1975-2000. Second national water assessment. Vol. 1: Summary. Superintendent of Documents, Washington, DC.

7. Weigner, K. K. 1979. The water crisis: it's almost here. Forbes, August 20, p. 56-63.

A Look at Turfgrass Water Conservation

Robert N. Carrow, Ph.D.
Professor, Crop and Soil Sciences Department
Georgia Experiment Station
University of Georgia
Griffin, GA

Introduction

Turfgrass water conservation has become increasingly important within the past 15 years because of
- *competition for scarce water resources*
- *water shortages during drought*
- *increased costs*
- *better awareness of turfgrass water use*

The goal is direct and simple: to minimize water use on turfgrass sites. Sometimes this goal is expressed in other terms, such as "turfgrass water conservation" or "efficient use of turfgrass water." Achievement, however, is complex, and each specific turfgrass site offers a unique set of challenges.

Water conservation under turfgrass situations has become increasingly important within the past 15 years. Numerous factors have brought this about, including increasing competition for scarce water resources, water shortages from periodic drought, increasing costs associated with obtaining water, and a greater awareness by turf managers that excessive irrigation is not necessary to maintain a good quality turf and often leads to other management problems.

Components of a Water Conservation Plan

Strategy is the science and art of devising and implementing plans or methods toward a goal. Turfgrass managers have nine methods that can be incorporated into an overall turfgrass water conservation strategy. Optimum water use efficiency cannot be achieved by any one approach but can with appropriate consideration of all nine methods. These nine components are: (1) develop/use grasses with high drought resistance, (2) develop/use grasses with high salt resistance, (3) use effluent or wastewater,

(4) alter management practices to efficiently use water, (5) improve irrigation system design, (6) improve irrigation scheduling, (7) use water harvesting, (8) provide continuing education, and (9) develop water conservation and drought contingency plans.

The United States Golf Association (USGA) has provided funding over the last ten years targeted toward research in most of these nine component areas. As a result, scientific and practical knowledge has greatly increased. Much attention has been directed to providing turf managers with new information in a timely manner[22]. Also, several scientific review articles have been published on various aspects of turfgrass water conservation namely by Gibeault and Cockerham[18], Carrow et al.[11], Kneebone et al.[24], and Balogh and Watson[3].

Develop/Use Grasses With High Drought Resistance

A primary, but long–term approach to decreasing water use is through the development of grasses with drought resistance. This requires (a) development of improved germplasm of turfgrass species, which are in current use, that possess greater drought resistance than current cultivars, including both introduced and native species, and (b) identification and improvement of grass species not currently used (or used to a very limited extent), but having potential as drought resistant turfgrasses.

Plant breeding and genetics obviously plays a dominant role in this approach. Breeders make plant explorations to obtain new genetic material. New selections are evaluated for quality, use, and drought resistance. Many times, some of the initial selections can be released as improved grasses, while using the various selections as breeding stock for even better second or third generation cultivars.

Unfortunately drought resistance is the most complex of all environmental stresses. Levitt[25] defined drought resistance as various morphological, anatomical, and physiological characteristics that a plant may have to withstand in periods of drought. Drought resistance consists of two major aspects: (a) drought avoidance (Table 1) is the

Nine methods to optimize water use efficiency:

- *develop/use grasses with better drought resistance*
- *develop/use grasses with salt resistance*
- *use wastewater*
- *alter management practices*
- *improve irrigation systems*
- *improve irrigation scheduling*
- *use water harvesting*
- *develop water conservation and drought contingency plans*

Drought resistance consists of various morphological, anatomical, and physiological characteristics that a plant may have to withstand in periods of drought.

25

ability of a plant to delay or postpone tissue dehydration, and, (b) drought tolerance (Table 2) is the ability of a plant to tolerate a drought period when tissues are exposed to dehydration.

Most plant characteristics listed in Tables 1 and 2 are present in the plant with or without a drought stress. Some characteristics develop (or are expressed to a greater degree) as a stress is induced such as thicker cuticles, wax in stomata, leaf senescence, smaller leaves, and all drought tolerance aspects. Regardless of whether they are present prior to a drought stress or develop during a stress, these factors contribute to overall drought resistance. Plants may exhibit a number of other responses to drought which are not beneficial. For example, photosynthesis declines as drought stress increases.

Each plant characteristic listed in Tables 1 and 2 influences overall drought resistance, is variable within a species, and is genetically determined. Breeders and plant physiologists work together to identify those factors that are most important within a species for influencing drought resistance. Beard and Engelke[4] reviewed this aspect and noted research knowledge is limited in understanding the most important mechanisms within each species that control drought resistance. At the intraspecific level, however, information is even more limited on most mechanisms, especially rooting depth/extent, and drought tolerance mechanisms. Data on overall drought avoidance and tolerance does exist at the intraspecies level, but the specific mechanisms for differences have not been well documented[4].

Breeders and plant physiologists work together to identify which plant characteristics are most important within a species for influencing drought resistance.

One important area that has often been neglected by breeders is the aspect of root tolerance to soil environmental stresses. Just as rooting depth/extent is genetically controlled, so is root tolerance. In most field situations, turfgrasses do not reach their genetic potential for rooting depth/extent because of intolerance to an adverse soil environment (i.e., soil strength, low O_2, high temperatures, Al/Mn toxicity, salinity, pests). Duncan[13] and Duncan and Shuman[15] noted this omission and incorporated it as a primary screening procedure to identify drought resistant germplasm.

One important area that has often been neglected by breeders is the aspect of root tolerance to soil environmental stresses.

Table 1. Drought Avoidance: Morphological, Anatomical, and Physiological Characteristics Contributing to Drought Avoidance

Drought avoidance is the ability of a plant to avoid tissue damage in a drought period by postponement of dehydration. The plant avoids dehydration by (a) exploring a greater volume of soil for water, and/or (b) reducing ET losses. Mechanisms involved are:

A. Development and maintenance of deep, extensive root system
   1. Greater root system size
      • deep rooting depth
      • high rate of root extension
      • high root length density
      • high number of root hairs
   2. Greater root tolerance (genetic) to soil stresses
      • high soil strength
      • low soil O_2
      • high soil temperatures
      • Al/Mn
      • high salinity and specific ions in salt affected soils
      • soil pests (insects, diseases)

B. Reduction of water use (evapotranspiration, ET)
   1. Important characteristics in moist soil that reduce ET
      • high canopy resistance
      • horizontal leaf orientation
      • high shoot density/verdure
      • low stomatal densities
      • hairy leaf surfaces
      • slow leaf growth rate
      • small conducting tissues
      • small/compact mesophyll cells in leaves
   2. Important characteristics in dry soils that reduce ET
      • stomatal closure
      • stomatal location
      • rolling, folding of leaves
      • thick, waxy cuticle
      • wax in stomata opening
      • leaf senescence
      • smaller leaves/low leaf area

Table 2. Drought Tolerance: Morphological, Anatomical, and Physiological Characteristics Contributing to Drought Tolerance

---

Drought tolerance is the ability of a turfgrass plant to tolerate a drought period of tissue dehydration. Potential mechanisms are:

A. Osmotic adjustment. The plant is able to make osmotic adjustments by accumulation of solutes to maintain turgor pressure, which is necessary to continue physiological processes.

B. Hardiness. A plant can possess greater drought hardiness (tolerance) due to:
   • greater membrane stability under dehydration
   • tolerance of protoplasmic constituents to dehydration
   • binding of cell water to protoplasm proteins and carbohydrates
   • loss of excess water contributing to succulence of tissues
   • accumulation of certain metabolites (proline, betaine, abscisic acid, etc.) may be significant in some instances

C. Escape. Some plants may escape the drought by living through the drought period:
   • as a seed
   • in a dormant state

The ultimate goals of plant breeders and plant physiologists are (a) to identify within a species the most important mechanisms contributing to drought resistance, and (b) develop rapid, reliable techniques to screen germplasm for these characteristics[4]. Care must be exercised to concentrate on all aspects of drought resistance mechanisms. Initially, plant factors contributing to low evapotranspiration (ET) under well–irrigated conditions received the most attention. In arid regions, where a high percentage of the water used by the turf must be applied, relative ET differences among cultivars to maintain a green turf have a reasonable correlation to overall drought resistance. In semi–arid and humid climates, however, periodic rainfall occurs and irrigation is used to supplement natural precipitation. For these conditions of erratic drought stress periods, ET under well–irrigated conditions provides no information on rooting aspects of drought avoidance or drought tolerance differences. Thus, it is important to identify important drought resistance mechanisms for a species under well irrigated *and* drought stressed (limited soil moisture) conditions. Table 1 identifies plant factors that influence ET under moist versus dry soil conditions.

Evapotranspiration (ET) correlates well with overall drought resistance in arid regions; however, in semi–arid and humid regions, ET is not well correlated with drought resistance.

Develop/Use Grasses With High Salt Resistance

Breeders must develop turfgrasses that can tolerate high soil salt levels and poorer water quality. These grasses would be used with lower water quality, effluent water, and on saltwater intrusion areas. Research by breeders, physiologists, and soil scientists on these problems is greatly hampered by location of the experimental site. Only a few of the current research facilities have soils with high salt content or poor water quality from ground water, effluent, or saltwater sources. Locations away from the established research facility are often not desirable because of the intensive management of turf and the necessity to obtain research data frequently. Problems associated with saline/sodic soils and poor water quality will increase in the future, as more effluent or wastewater is applied to turf sites. Recent reviews concerning salinity

Problems associated with saline/sodic soils and poor water quality will increase in the future, as more effluent or wastewater is applied to turf sites.

and turfgrass breeding and culture have been written by Kenna and Horst[23] and Harivandi et al.[19]

Problems with salt affected soils:
- *water availability*
- *poor soil physical conditions (massive structure)*
- *ion toxicities*
- *nutrient imbalances*

Salt–affected soils may exhibit several problems: (a) high osmotic potential which limits plant water availability, (b) poor soil physical conditions are common on sodic soils due to deflocculation of soil colloids by Na, (c) ion toxicities may occur, and (d) nutrient imbalances may be present[17]. These soil problems can induce many adverse plant responses (Table 3). Turfgrasses may exhibit several salt resistance mechanisms to minimize salt stress (Table 4).

Interspecies salt resistance differences are reasonably well defined and some information exists at the intraspecies level[19,23]. Very limited research, however, has been directed to identifying the specific avoidance or tolerance mechanisms associated with salt resistance of any particular species. As with drought resistance, once important mechanisms inferring salt resistance are identified for a species, breeders or plant physiologists may develop simple screening procedures for germplasm.

Very limited research has been directed to identifying the specific avoidance or tolerance mechanisms associated with salt resistance of any particular species.

The alternative would be to screen all germplasm in a controlled environment using a "standard" soil to impose the stresses. If this approach is taken, a saline/sodic soil would offer two advantages. First, soil allows dehydration of tissues (especially root tissues), while solution culture does not. Thus, some important plant responses may not occur in a solution culture but could in a soil; namely, reduced root cell permeability, reduced root cell metabolic rate, disruption of root cell protoplasm, and root elongation may be more affected. Second, a saline/sodic soil has high levels of total salts and Na but does not have the poor soil physical properties of a sodic soil. Sodium is the one specific ion most likely to be toxic at high levels in salt affected soils. Since the second most common ion that may cause ion toxicity is Cl, the saline/sodic should also contain high Cl levels. A saline/sodic soil with a sandy loam or loamy sand texture could possess adequate physical properties for use in a greenhouse screening procedure. With some experience, a good saline/sodic soil could be identified to provide a screening standard for the major salt–related stresses.

Table 3. Salinity Effects on Turfgrass[a]

Effect	Tissue
• slow/stop cell elongation	R[c]
• hasten maturity/suberization of root cells	R
• reduce permeability of root cells	R[b,c]
• enhance Casparian strip formation in root cells	R
• reduce metabolic activity	R/S[b,c]
• disrupt protoplasm	R/S[b,c]
• induce hormonal imbalances	R/S
• acute leaf tip or margin burn	S[b]
• specific ion effects on anatomy or physiology	R/S

[a] After Follett et al.[17], Duncan[14], and Hoffman[21].
[b] Due to tissue dehydration.
[c] Whether a dry soil or a moist environment (hydroponic) is present influences the relative importance of "effects". Items noted with a "c" are less likely to appear in a moist, salinity situation versus a dry soil, salinity condition.
R = roots, S = shoots

Table 4. Salt Resistance: Mechanisms in Turfgrass

Salt resistance is the ability of a turfgrass to withstand salt stress. Resistance may be due to salt avoidance and/or tolerance.

A. Salt avoidance. Mechanisms by which the plant avoids or delays onset of salt stress
   • selectively limit total salt uptake
   • selectively limit uptake of a specific ion (such as Na^+, Cl^-, etc.)
   • secretion of salts

B. Salt tolerance. Mechanisms allowing the plant to tolerate high stress.
   • tissue dehydration tolerance
   • osmotic adjustment to maintain turgor pressure
   • tissue tolerance to specific ions

Duncan[14] recently reviewed the current status of genetic manipulation of edaphic soil related stresses, including salt stress. He noted the importance of developing standardized procedures to induce edaphic stresses that are highly representative of the actual stress on the plant in a field situation. In the above example, a saline/sodic soil with high salt, Na, and Cl levels could induce root tissue dehydration during drying cycles. This would appear to be a more sensitive approach than use of a hydroponic system, regardless of total salt, Na, and Cl concentrations. Duncan[14] also noted the importance of imposing severe stress levels (i.e., selection index of 1% to 5% survival of germplasm) to identify truly resistant germplasm to a particular edaphic stress.

"Ideal" turfgrass for water conservation:
- *high drought avoidance and tolerance*
- *high salt avoidance and tolerance*
- *tolerance of environmental, traffic, and pest stresses*

Combining the information on drought and salt resistance, the "ideal" turfgrass for water conservation should possess (a) high drought avoidance and tolerance, (b) high salt avoidance and tolerance, and (c) good tolerances to other environmental, traffic, and pest stresses, while producing a quality turfgrass. At first glance, this appears "idealistic", but significant gains toward this goal have been achieved over the past decade.

Use of Effluent/Wastewater

A third strategy closely associated with the previous one is the use of effluent water. This has been a common practice in arid regions. In the future, this practice will increasingly spread to humid regions, especially in urban settings where potable water is at a premium. As this occurs, problems will develop that may not appear in arid or semi–arid regions. Refinements in cultural practices will be required to ensure efficient, safe use of effluent water.

Since the remaining chapters deal with various aspects of using wastewater in turfgrass situations, a detailed discussion of this topic will be omitted. As background material, the 1978 proceedings of Cleaver et al.[12] and the more recent (1985) book by Pettygrove and Asano[29] are suggested.

Alter Management Practices To Efficiently Use Water

Plant morphological, anatomical, and physiological characteristics that increase drought/salt avoidance and drought/ salt tolerance are genetically determined with variation between species and among cultivars of a species; but they are also influenced by cultural practices, environmental (soil and air) conditions, and turf use (wear, soil compaction). In addition to directly affecting plant drought/ salt resistance characteristics, cultural practices may (a) alter soil water balance by influencing runoff, evaporation, transpiration, soil water retention, and leaching, and (b) influence salt problems such as poor structure, high salinity, high sodium, or specific ion toxicities. Thus, cultural practices have a profound influence on turfgrass water relations. They have the capability of improving conditions so the plant can be near its genetic capability in drought/salt resistance, or, conversely, improper cultural practices can negate the plant's inherent genetic potential to resist drought/salt stresses.

Cultural practices help alter soil
* *water balance*
* *structure*
* *salinity and sodium levels*
* *toxic ion levels*

Current status on how management practices influence turfgrass water relations have been reviewed by Carrow et al.[11], Kneebone et al.[24], and Balogh and Watson[3] (Table 5). The role of soil physical conditions on water conservation aspects was presented by Carrow[7] and Carrow and Petrovic[10]. Harivandi et al.[19] reviewed cultural practices to alleviate salinity problems and enhance water use efficiency. Carrow[8] has provided a practical guide to maintaining maximum root depth and viability for water conservation. As more basic knowledge evolves about how plant and soil aspects are altered by individual or combinations of cultural practices, we can develop much better cultural regimes. To be most effective, this will need to be done at the cultivar level, because within species, cultivars may vary substantially to cultural practices.

To be most effective, cultural practice regimes need to be done at the cultivar level because of the substantial variation which occurs within species.

Improved Irrigation System Design

Irrigation system design is an important approach to water conservation. Essential features that improve water use by irrigation design are:

Table 5. Turfgrass Cultural Aspects That Influence Water Relationships

Cultural Aspect[a]	Relative Importance [b]
	1=most, 3=least
Selection species	1
Selection of cultivar	1
Turfgrass use and quality expectation	1
Mowing height	1
Nitrogen fertilization	1
Soil cultivation	1
Soil modification	1
Irrigation time of day	1
Irrigation rate and frequency	1
Alleviation of salt affected soils	1
Liming acid soils	1-2
Mowing frequency	2
Potassium fertilization	2
Mowing blade sharpness	2-3
Plant growth regulators	2-3
Wetting agents	2-3
Soil insect control	2-3
Soil disease control	2-3
Phosphorus nutrition	3
Iron nutrition	3
Antitranspirants	3
Other pesticides with PGR activity	3

[a] From Carrow et al.[11], Kneebone et al.[24], and Balogh and Watson[3].

[b] Author's ranking based on literature and experience. On any particular site, the relative importance may change; for example, a serious grub infestation would require immediate insect control in order to maintain turf roots for water uptake.

- designing for uniformity of application to minimize wet and dry areas

- zoning irrigation heads of similar areas together

- using fewer irrigation heads per zone

- matching application rate to soil infiltration rate

- use of low volume heads when appropriate

- use of multiple water applications by pulse irrigation to allow infiltration

- controller flexibility to develop the most efficient irrigation program

- elimination of pipe leakage

Irrigation system design is an important approach to water conservation.

A detailed discussion of irrigation system design for water conservation has been provided by Olson[27] and Meyer and Camenga[26]. Wade et al.[32] provide a practical discussion of the role of irrigation design as a component in the xeriscape concept.

Improved Irrigation Scheduling

A sixth approach is by improved irrigation scheduling. Technology is developing rapidly in this area and can assist the grower in reducing water runoff, leaching, and excess evaporation losses. Technological tools assisting turf managers in irrigation decisions are (a) soil–based, to monitor soil water status; (b) plant–based, to monitor plant water status; and (c) atmospheric, to monitor atmospheric conditions that influence ET rates.

Irrigation scheduling can reduce
- *runoff*
- *leaching*
- *evaporation losses*

Examples of soil–based tools are an increasing array of soil moisture sensors in addition to tensiometers and moisture resistance blocks that have been available for many years. Often these sensors work on different principles than tensiometers or moisture blocks and may not have the limitations of these instruments.

Three irrigation scheduling methods:
- *soil–based*
- *plant–based*
- *atmosphere–based*

The oldest plant–based irrigation guide is observation for wilt, but being able to determine stress before visual wilt symptoms would be very beneficial. Systems that monitor canopy temperatures are now available and can be used to help schedule irrigation.

Examples of atmospheric–based tools are weather pan evaporation and estimating ET by various environmental–based formulas (Penman equation, others). State–of–the–art irrigation controllers, coupled with weather–monitoring devices, are now available from major irrigation manufacturers. Further improvements in atmospheric–based approaches can be expected as the data base builds and growers begin to use the full capabilities of these systems.

State–of–the–art irrigation controllers, coupled with weather–monitoring devices, are now available from major irrigation manufacturers.

Kneebone et al.[24] summarized the current status of irrigation scheduling technology on turfgrass, while Carrow[9] recently discussed use of canopy temperatures for irrigation scheduling. Good technical discussions on this topic across all crops can be found in the articles by Burman et al.[6] and Heermann et al.[20] .

One aspect that has not received adequate research attention is irrigation scheduling on salt–affected turfgrass soils for their reclamation. Within arid regions, irrigation scheduling by ET data or other methods in order to conserve water is often stressed. On salt–affected soils, however, this is counter–productive if salts cannot be leached. The best long–term water conservation approach on salt–affected soils is to reduce the salt problems by leaching. Technology for scheduling irrigation can still be used but the extra water required to leach salts must be incorporated into the turf manager's water conservation plan. Otherwise, salt–affected soils may become increasingly common.

The best long–term water conservation approach on salt–affected soils is to reduce the salt problems by leaching.

Ironically, recycled water may contribute to salt buildup problems, while at the same time, be the best solution to salinity problems.

Ironically, recycled water may contribute to salt buildup problems while at the same time, be the best solution to salinity problems. Leaching of salts requires extra water beyond normal ET levels. Recycled water can be the source of this "extra" quantity of water. Tile lines to collect the leaching fraction and for removal of leached salt are highly desired.

Additional research is needed as to the appropriate combination of leaching, chemical amendments (where needed) to offset soil and/or irrigation water quality deficiencies, cultivation, and tile drainage under turfgrass situations. As an example, gypsum can be applied through turf irrigation systems on sodic soils to counterbalance high Na in soil and possibly in irrigation water. Cultivation techniques are also available to apply gypsum in vertical columns to a depth of 30 cm with 15 cm spacing and as bands at 18 cm depth and 13 cm spacings. It is possible that a combination of surface treatment with gypsum plus localized vertical and horizontal bands could reclaim sodic soils more effectively than any one approach alone, thereby enhancing water use efficiency.

Water harvesting is thought of as treating watersheds to enhance runoff which is collected for future use. A broader definition could include use of storage ponds to collect surface water from streams or subsurface water for future use.

Oster and Rhoades[28] discuss water management on salt-affected soils, including use of wastewater. Rhoades and Loveday[30] also provide a detailed review of salinity in irrigated agriculture.

Water Harvesting

On some sites, water harvesting may be a feasible conservation measure. Normally, water harvesting is thought of as treating watersheds to enhance runoff which is collected for future use. A broader definition could include use of storage ponds to collect surface water from streams or subsurface water for future use. Also, recycling water may be achievable for some sites. Bertrand[5] reports on different approaches to water harvesting.

Education

Individual turf managers will be increasingly challenged by more sophisticated technology (i.e., infrared thermometers for canopy temperatures and complex controllers with tremendous ability to provide detailed information); the need to refine management programs to smaller and smaller units (green–by–green, tee–by–tee), pressure to incorporate all possible water conservation tactics, expectations to provide data on the degree of water saved, the necessity to alter many management practices for a specific cultivar in order to achieve maximum benefits, and other similar challenges.

Turf managers will be challenged by
- *more sophisticated technology*
- *the need to refine management programs*
- *pressure to conserve water*
- *need for records showing water savings*
- *need to alter management practices*

37

An essential step in an overall strategy to conserve water on turfgrass is development of specific water conservation and drought contingency plans at all levels.

To their base of knowledge, managers will need to understand new technological and scientific advances in turfgrass science in order to intelligently incorporate these into their management schemes. At the industry level, the USGA Green Section, GCSAA, manufacturers, and university personnel must find efficient ways to transmit the latest knowledge rapidly and in a comprehensive, understandable manner. Many of the references in this article, and the article itself, are attempts to summarize and organize information on turfgrass water conservation into an understandable format.

Water Conservation and Contingency Plans

An essential step in an overall strategy to conserve water on turfgrass is development of specific water conservation and drought contingency plans at all levels—specific turf site, city/county, water district, and state. A review by the turf manager of the previous strategies will reveal how water conservation measures can be incorporated into such plans. Just as with pesticide issues confronting the lawn–care industry, the most successful tactic has been input into regulatory and governing agencies before regulations are developed.

A water conservation plan is targeted to conserving water on a continuous basis, while a contingency plan deals with water conserving measures in times of water shortage.

A water conservation plan is targeted to conserving water on a continuous basis, while a contingency plan deals with water conserving measures in times of water shortage. Many times both are incorporated into an overall plan. Agnew and Carrow[2], Adams[1], and Smith[31] discuss development of conservation/contingency plans for a specific site, within a state, and on an interstate (regional) basis, respectively. Water conservation on turfgrass should be just one component of a water conservation plan for any unit—whether a single golf course, city, region, or state. All water users (industrial, commercial, residential, turf, agricultural, public uses, public system losses, etc.) must be factored together. Flack et al.[16] illustrate this concept. While they do not incorporate all aspects, they, at least, discuss non–residential water consumers.

Summary

A sound strategy for turfgrass water conservation cannot rely on just one component, use of wastewater for example. Instead, turfgrass managers must consider all components of the "turfgrass system" and integrate all aspects that are feasible for their unique conditions. Nine components of a good water conservation plan are:

Turfgrass managers must consider all components of the "turfgrass system" and integrate all aspects that are feasible for their unique conditions.

- use of grasses with high drought resistance

- use of grasses with high salt resistance

- altering management practices to efficiently use water

- a well–designed irrigation system

- improved irrigation scheduling

- water harvesting

- continuous growth in knowledge related to water issues

- development and implementation of water conservation and drought contingency plans

Considerable progress has been achieved over the past decade in most of the above areas in terms of research and practical information. This trend will continue but only if we go beyond our current state of knowledge.

References

1. Adams, B. 1981. Responsibility during periods of water shortage: The role of the South Florida Water Management District. p. 28-32. *In* Proc. 29th Florida Turf. Conf., Univ. of Florida, Gainesville, FL.

2. Agnew, M. L. and R. N. Carrow. 1986. Managing water for turfgrass. Grounds Maintenance 21(3):90-121.

3. Balogh, J. C. and J. R. Watson. 1992. Role and conservation of water resources. Chap. 2. In J.

C. Balogh and J. W. Walker (ed.) Golf course management and construction:environmental issues. Lewis Publishers, Boca Raton, FL.

4.   Beard, J. B. and M. C. Engelke. 1985. An environmental genetics model for turfgrass improvement: physiological aspects. Chap. 1. *In* F. LeMaire (ed.) Proc. 5th Int'l. Turfgrass Research Conf. Institute Nat. de la Recherche Agron., Paris, France.

5.   Bertrand, A. R. 1966. Water conservation through improved practices. p. 207-235. W. H. Pierre et al. (ed.) Plant environment and efficient water use. American Society of Agronomy, Madison, WI.

6.   Burman, R. D., R. H. Cuenca, and A. Weiss. 1983. Techniques for estimating irrigation water requirements. p. 273-335. *In* D. Hillel (ed.) Advances in irrigation, Vol. 2, Academic Press, NY.

7.   Carrow, R. N. 1985. Soil/water relationships in turfgrass. Chap. 8. *In* V. A. Gibeault and S. T. Cockerham. (ed.) Turfgrass water conservation. Pub. 21405. Coop. Ext. Service, Univ. of California, Oakland, CA.

8.   Carrow, R. N. 1989. Managing turf for maximum root growth. Golf Course Management 57(7):18-26.

9.   Carrow, R. N. 1993. Canopy temperature irrigation scheduling indices for turfgrasses in humid climates. Int'l. Turfgrass Soc. Research Journal. 7:594-599.

10. Carrow, R. N. and A. M. Petrovic. 1992. Effects of traffic on turfgrass. Chap. 9. *In* D. V. Waddington, R. N. Carrow, and R. C. Shearman (ed.) Turfgrass, Monograph No. 32. American Society of Agronomy, Madison, WI.

11. Carrow, R. N., R. C. Shearman, and J. R. Watson. 1990. Turfgrass. Chap. 30. *In* B. A. Stewart and D. R. Nielsen (ed.) Irrigation of Agricultural

Crops, Monograph No. 30, American Society of Agronomy, Madison, WI.

12.   Cleaver, G. W., H. M. Ferguson, A. M. Radko, and D. A. Rossi. (ed.)1978. Proc. of Wastewater Conference. Golf Course Supt. Assoc. of America and USGA. Far Hills, NJ.

13.   Duncan, R. R. 1991. Acid soil tolerance breeding in sorghum. Adv. Agron. (India) 1:71-79.

14.   Duncan, R. R. 1993. Genetic manipulation. Chap. 7. *In* R. E. Wilkinson (ed.) Plant response mechanisms to the environment. Marcel Dekker, NY (in press).

15.   Duncan, R. R. and L. M. Shuman. 1993. Acid soil stress response of zoysiagrass. Int'l. Turfgrass Soc. Research Journal. 7:805-811.

16.   Flack, J. E., W. P. Weakley, and D. W. Hill. 1977. Achieving urban water conservation: a handbook. Colorado Water Resources Research Institute, Completion Report 80. Colorado State University, Fort Collins, CO.

17.   Follett, R. N., L. S. Murphy, and R. L. Donohue. 1981. Fertilizers and soil amendments, Chap. 9. Prentice-Hall Inc., Englewood Cliffs, NJ.

18.   Gibeault, V. A. and S. T. Cockerham (ed.) 1985. Turfgrass water conservation. Publication No. 21465. Coop. Ext. Service, Univ. of California, Oakland, CA.

19.   Harivandi, M. A., J. D. Butler, and L. Wu. 1992. Salinity and turfgrass culture. Chap. 6. *In* D. V. Waddington, R. N. Carrow, and R. C. Shearman (ed.) Turfgrass, Monograph No. 32. American Society of Agronomy, Madison, WI.

20.   Heermann, D. F., D. L. Martin, R. D. Jackson, and E. C. Stegman. 1990. Irrigation scheduling controls and techniques. Chap. 17. *In* B. A. Stewart and D. R. Nielsen (ed.) Irrigation of Agricultural Crops, Monograph No. 30. American Society of Agronomy, Madison, WI.

21.   Hoffman, G. J. 1981. Alleviating salinity stress.

Chap. 9. *In* G. F. Arkin and H. M. Taylor (ed.) Modifying the root environment to reduce crop stress. American Society of Agricultural Engineers, St. Joseph, MO.

22. Kenna, M. P. 1992. USGA/GCSAA research results you can use. USGA Green Section Record 30(3):6-9.

23. Kenna, M. P. and G. L. Horst. 1993. Turfgrass water conservation and quality. Int'l. Turfgrass Soc. Research Journal. 7:99-113.

24. Kneebone, W. R., D. M. Kopec, and C. F. Mancino. 1992. Water requirements and irrigation. Chap. 12. *In* D. V. Waddington, R. N. Carrow, and R. C. Shearman (ed.) Turfgrass. Monograph No. 32. American Society of Agronomy, Madison, WI.

25. Levitt, J. 1980. Responses of plants to environmental stresses. Vol. II. Academic Press, NY.

26. Meyer, J. L. and B. C. Camenga. 1985. Irrigation systems for water conservation. Chap. 9. *In* V. A. Gibeault and S. T. Cockerham (ed.) Turfgrass Water Conservation. Publication No. 21405. Coop. Ext. Service. Univ. of California, Oakland, CA.

27. Olson, C. O. 1985. Site design for water conservation. Chap. 11. *In* V. A. Gibeault and S. T. Cockerham (ed.). Turfgrass water conservation. Publication No. 21405. Cooperative Extension Service, University of California, Oakland, CA.

28. Oster, J. B. and J. D. Rhoades. 1985. Water management for salinity and sodicity control. Chap. 7. *In* G. S. Pettygrove and T. Asano, (ed.) Irrigation with reclaimed municipal wastewater —a guidance manual. Lewis Publications, Chelsea, MI.

29. Pettygrove, G. S. and T. Asano (ed.) 1985. Irrigation with reclaimed municipal wastewater —a guidance manual. Lewis Publications, Chelsea, MI.

30. Rhoades, J. D. and J. Loveday. 1990. Salinity in irrigated agriculture. Chap. 36. *In* B. A. Stewart and R. D. Nielsen (ed.) Irrigation of Agricultural Crops. Monograph No. 32. American Society of Agronomy, Madison, WI.

31. Smith, M. R. 1985. Drought emergency planning. USGA Green Section Record 23(3):11-13.

32. Wade, G. L., J. T. Midcap, K. D. Coder, G. Landry, A. W. Tyson, and N. Weatherby, Jr. 1992. Xeriscape—a guide to developing a water–wise landscape. Coop. Ext. Service Bulletin 1073. Univ. of Georgia, Athens, GA.

Effluent for Irrigation: Wave of the Future?

Garrett Gill and
David Rainville
Members
American Society of
Golf Course Architects

Introduction

Beginning in the mid–sixties, the use of effluent or wastewater on golf courses was thought to be the answer to many developers' dilemmas in states with restrictive water use laws such as Arizona and California. The thought and attitude at that time was that effluent was free water and free was great.

Effluent or wastewater on golf courses was thought to be the answer to many developers' dilemmas in states with restrictive water use laws.

With regard to "Effluent for irrigation, the wave of the future," effluent has been present in the golf industry for more than 25 years. The majority of its use has been in the arid regions of the U.S. The wave of the future should focus on extending wastewater use beyond the current use boundaries and beyond the few case studies in humid regions.

The whole notion of reusing, recycling, and conservation of resources was, and continues to be, very popular with the general public and among state and local agencies. The concept that a large scale residential golf community and the golf course itself could be water interdependent was a major asset to those types of developments.

The whole notion of reusing, recycling, and conservation of resources was, and continues to be, very popular with the general public and among state and local agencies.

From a golf course architect's perspective, the use of effluent has had varying degrees of impact. Golf course architectural firms with projects on the West Coast and in the Southwest have relied heavily on the use of effluent in their designs or to support their designs. As one looks at golf course projects in other regions, wastewater as the prime supply source drops off, though its use as a secondary or back–up source remains high.

Golf course architects view the use of effluent water with mixed feelings. Often a golf course designer wishes that a feasible alternative fresh water source was available, but that is seldom the case in the water–poor areas of the world.

From a design perspective, the use of effluent has some effect. Obviously, courses are not longer, shorter, wider or narrower with or without the use of effluent. However, to accommodate the use of effluent and the impacts associated with its use, golf courses are turfed, landscaped and irrigated differently than traditional fresh water courses.

To accommodate the use of effluent and the impacts associated with its use, golf courses are turfed, landscaped and irrigated differently than traditional fresh water courses.

This paper addresses, from a golf course architect's perspective, the issues of the expanding application of wastewater for golf course irrigation. These issues target wastewater use on golf courses and how its use affects costs, aesthetics, quality of play and saleability of the idea. At the end of this paper, a checklist is presented which provides significant points for developers and operators regarding the use of effluent on their golf course projects.

Issues Surrounding Wastewater Use on Golf Courses

A majority of locations where golf is played require some degree of supplemental water to provide adequate and durable turf cover. Therefore, the source of irrigation water supply for the course is usually a factor.

A common misconception is that wastewater as a golf course irrigation source is only applicable in the water–poor, semi–arid or arid environments.

A common misconception is that wastewater as a golf course irrigation source is only applicable in the water–poor, semi–arid or arid environments. Most water–poor communities are reluctant to add amenities and facilities that have high potable water demands. In order to build golf courses in these communities, the compromise is to use more drought tolerant plant materials and develop alternative water use schemes. As more water–poor communities have successfully used effluent for irrigation purposes, techniques and water management solutions have been developed sufficiently so that the technology transfer of effluent use can be made to other areas, including humid environments.

As more communities are faced with increased water supply costs, watershed protection plans, and public concerns about water usage, effluent use can provide a recycling solution that is often acceptable to the public.

As more communities are faced with increased water supply costs, watershed protection plans, and public concerns about water usage, effluent use can provide a recycling solution that is often acceptable to the public. Effluent provides an answer to competing water demands in many communities. The following points are significant with respect to extending the application of wastewater beyond the water–poor, arid environments:

- adverse effects of wastewater usage are less apparent because of dilution and leaching by natural rainfall

- communities can evaluate the positive use of wastewater on their golf courses as a means of reducing the demand upon existing groundwater sources and upon existing water supply systems

- communities can also examine the use of wastewater on the basis of "we have it, let's use it" and on the basis of cost recovery for the cost of treatment and distribution of effluent.

Case studies have shown that effluent can be safely used on both golf courses and within housing/golf course developments.

Case studies have shown that effluent can be safely used on both golf courses and within housing/golf course developments. Although the requirements vary from state to state and with the degree of treatment, golf courses are using effluent for irrigation in almost every state.

The golf industry continues to adapt and adjust to the issue of water. The irrigation industry has made practically all of their products contamination resistant. Special colored pipes and valve markers now clearly indicate wastewater lines. The turf industry has continued to develop improved strains of salt tolerant, and drought tolerant turfgrasses for golf courses.

The irrigation industry has made practically all of their products contamination resistant.

In response to limitations on water use, such as in Arizona, the industry has created, and the golfing public accepted, the desert style golf course. This type of successful innovation has been adapted to east coast, gulf coast, inter–mountain, mountain, south and midwest areas, though systems in these regions are not as noticeable or distinctive in their effluent plans.

Issues of Cost

Wastewater use on golf courses is not free in any sense. In California, golf courses are required to connect to a wastewater line if it is available. That water is metered and charged at rates nearly equivalent to potable fresh water costs. In addition to the actual cost of the wastewater, golf courses using effluent will generally have greater maintenance costs. The increased costs are derived from

Wastewater use on golf courses is not free . . . In addition to the actual cost of the wastewater, golf courses using effluent will generally have greater maintenance costs.

- use of additional soil amendments to mitigate high salt and sodium levels resulting from effluent water use

- higher costs associated with developing fresh water leaching systems for greens

- higher water quality management costs (to keep BOD, TDS and TSS levels in check)

- increased use of fertilizers, herbicides and fungicides due to overall poorer turf quality

- increased costs associated with more advanced and area–specific irrigation system designs, including the use of drip and subsurface systems.

Issues of Aesthetics and the Quality of Play

The use of wastewater does impact aesthetics and quality of play, but not directly from the end user perspective. The question is best answered by the user, in this case the golfer. The golfer is not directly concerned with the type or brand of irrigation system in place, the types and species of turfgrass or trees present, or the type of water used in irrigation. In fact, besides the warning at the

The use of wastewater does impact aesthetics and quality of play, but not directly from the end user perspective.

proshop and signs by the irrigation pond, there are no other visual clues that wastewater is being used.

However, what the golfer does notice is course condition, most notably the condition of the greens, tees and fairways. Poor turf quality in these areas is usually blamed on the superintendent or course operator and not on the use of wastewater. Typically water quality is the last aspect to be investigated.

Issues of Being Saleable

One of the difficult points of effluent use is selling the concept to a careful public.

With respect to creating a saleable concept, how do we redefine the word "wastewater" to make it politically palatable and acceptable to the public? How do we sell the concept of effluent use to a public that envisions a potty pumper or honey wagon coming up to Hole #6 and spraying human waste and water on the fairways?

One of the difficult points of effluent use is selling the concept to a careful public. The general public is constantly being educated in what we term Environmental Consequences 101. Already golf courses may fall victim to public scrutiny because of the use of pesticides, fertilizers and the danger of human contact with these materials, groundwater contamination, groundwater depletion, wetland loss, and wetland degradation. It is easy to see how the public, in certain circumstances, can become wary of the concept of irrigating golf courses with wastewater. They may wish to impose additional restrictions and limitations that make the project too costly to construct.

Choice of words is important in public meetings—using terms such as "reclaimed water" or "effluent."

Just as we have had to rethink the meaning of garbage and trash, and separating out the recyclable materials, the same thought process is required for effluent use. A good explanation of how effluent actually reuses nutrients present in the water can help sell this concept. The large acreage that requires water and the fact that turf is not a food crop makes golf courses good candidates for effluent use. Choice of words is important in public meetings—using terms such as "reclaimed water" or "effluent" is better than saying wastewater.

The concept of using wastewater as a selling tool returns to the basic theme of conservation and reuse. To the municipality that has paid once or twice to clean the wastewater, charging for the use of the cleansed water to recoup some of the monies makes financial sense compared to discharging treated water directly to a waterbody. To the developer who may have to purchase water rights for the residential and commercial portions of the golf development, the reuse again of twice–paid–for water is a financial necessity.

To the developer who may have to purchase water rights for the residential and commercial portions of the golf develop-ment, the reuse again of twice–paid– for water is a financial necessity.

Developers/Operators Checklist for Use of Wastewater in a Golf Course Irrigation Program

Sampling Soils

1. Soil samples should be taken from different sec-tions of the golf course including tees, greens, fairways and rough. This should be done as much in advance of the water conversion as possible to allow for tracking of changes due to the conversion to wastewater.

2. Sample soils on a quarterly basis. Monitoring soils will allow for adjustments in the water schedule and any necessary mitigating measures.

Water Quality

1. Establish initial analysis to determine if further treatment is necessary. Water should be analyzed periodically, along with the soil.

2. Verify source of wastewater. Wastewater from industrial areas can have a greater amount of un-desirable elements as compared to wastewater from essentially residential areas.

Wastewater from industrial areas can have a greater amount of undesirable elements as compared to wastewater from essentially residential areas.

49

Verify the level of treatment. Advanced filtering is best and is one stage beyond tertiary treatment.

3. Verify the level of treatment. Advanced filtering is best and is one stage beyond tertiary treatment.

4. Negotiate with the supplying entity to establish maximum levels of biological oxygen demand (BOD), total dissolved salts (TDS), and total suspended solids (TSS). It is much easier for the end user to have the treatment plant control undesirable elements.

Pumping and Water Storage

1. If possible, draw water directly from a pressurized source pipe line and apply directly to the golf course.

2. If insufficient line pressure exists, install a booster pumping station to increase pressure to golf course levels.

If water must be stored, the first choice is an enclosed tank, eliminating exposure to sunlight and reducing the formation of algae.

3. If water must be stored, the first choice is an enclosed tank, eliminating exposure to sunlight and reducing the formation of algae. The second and most likely choice is storage in lakes. Keep to a minimum the number of lakes to reduce the problem of algae control.

4. The deeper the lake, the better. This helps reduce sunlight penetration, the water stays cooler, and algae control systems are more effective.

5. Blending wastewater with fresh water in ponds is usually helpful in controlling BOD's, TSD's and TSS's. Ozonation as one type of microfloculant is also helpful. Ozonation increases the amount of oxygen in the water and reduces surface tension allowing compounds to drop out of suspension [1]. Blending requires a separation of systems to prevent reclaimed water from contaminating a fresh water source.

6. A dual irrigation system is preferred if fresh water is available. Fresh water should be applied to greens, tees, ornamental lakes, and other sensitive plantings.

Miscellaneous

1. Drinking fountains need to have a self-closing cover. Check with local authorities.

2. Signs should indicate "reclaimed water used to irrigate turf."

3. If there is any possibility of reclaimed water being used in the future, design the irrigation system to be easily converted.

   a. Use warning tape or colored pipe depending on local codes.

   b. Design for the proper separation of domestic pipelines from reclaimed pipelines (usually ten foot horizontal and one foot vertical where lines cross).

4. Negotiate the hours of operation with the treatment plant. Some plants will operate at lower supply line pressure during daylight hours and boost pressure at night allowing for direct irrigation without boosting pumps.

5. Install low pressure sensors to shut down the system in the event of pressure failures.

6. Every possible operation should be automated, leaving nothing to chance.

7. Devise a backup system, at least for greens and tees. Some treatment plants shut down periodically for maintenance of their systems.

If there is any possibility of reclaimed water being used in the future, design the irrigation system to be easily converted.

Devise a backup system, at least for greens and tees. Some treatment plants shut down periodically for maintenance of their systems.

Summary

Our efforts, as golf course architects and as members of the American Society of Golf Course Architects, will focus on striving for improvements and advancements in the technological and social impacts of effluent use on golf courses. Specifically it is important that:

- the turf industry continues to perform due diligence towards research on new turf varieties that are "wastewater tolerant"

The turf industry should continue research developing "wastewater tolerant" varieties.

The use of wastewater as an irrigation source will continue to be a part of the golf development picture in the future.

- the irrigation industry continue its orientation toward conservation of all water resources

- research on the treatment and reuse of wastewater be continued with emphasis on cost effective methods to manage BOD, TDS and TSS levels

- there is increased education and public awareness about the use of wastewater as an irrigation source and wastewater's role in global water management programs.

"Effluent for irrigation, wave of the future" suggests that the use of wastewater as an irrigation source will continue to be a part of the golf development picture in the future.

References

1.   Green, S. and F. Gardner. 1992. Managing effluent water use in course maintenance. Golf Development Magazine. June 1992. p. 4-5.

Chapter 2   Regulations, Ordinances and Legal Liabilities

Regulations Affecting the Use of Wastewater on Golf Courses

James Crook, Ph.D., P.E.
Associate
Camp, Dresser & McKee Inc.

Water Rights: Legal Aspects and Legal Liability

Anne Townsend Thomas
Partner
Best, Best & Krieger

Regulations Affecting the Use of Wastewater on Golf Courses

James Crook, Ph.D.,P.E.
Associate
Camp, Dresser & McKee Inc.

Introduction

The beneficial reuse of municipal wastewater is an integral component of water resources management in many parts of the United States and the use of reclaimed water for golf course irrigation is well–established in many states. There are more than 150 golf courses irrigated with reclaimed water in California and Florida alone [4,11]. As droughts and population increases continue to stress freshwater supplies, reuse of municipal wastewater will play an ever–increasing role in helping to meet water demands.

There are more than 150 golf courses irrigated with reclaimed water in California and Florida alone.

The applicability of reclaimed water for golf course irrigation depends on its physical, chemical and microbiological quality. The effects of physical parameters and chemical constituents are, for the most part, well–understood, and recommended criteria have been established by the U.S. Environmental Protection Agency (EPA) and others. Health–related microbiological limits are more difficult to quantify, as evidenced by widely varying state standards and guidelines. There are no federal standards governing water reuse in the U.S., and regulations that do exist have been developed at the state level.

The applicability of reclaimed water for golf course irrigation depends on its physical, chemical and microbiological quality.

Northern and midwestern states with ample water resources typically have no water reuse projects or regulations, and many other states have few reuse projects and minimal regulations. As might be expected, in water–short states such as Arizona, California, Florida, and Texas where reuse is practiced extensively, well–developed, comprehensive water reclamation and reuse regulations have been established. The EPA recently published water reuse guidelines that are intended to provide guidance to states that have not developed their own criteria or guidelines.

Rationale for Reuse

The planned reuse of municipal wastewater for nonpotable purposes, including golf course irrigation, and for indirect augmentation of potable water supplies has been practiced for many years throughout the world. With the concomitant increase in water demand as population grows, purposeful water reuse will play an increasing role in the planning and development of additional water supplies. Water reuse reduces the demand on freshwater supplies through "source substitution," that is substituting reclaimed water for potable or other water for golf course irrigation and other uses where the quality of the reclaimed water permits such substitution.

Reasons for reusing wastewater:
- *opportunity*
- *need*
- *conservation*
- *reliability of supply*
- *well–established technology*
- *economics*
- *pollution abatement*
- *public policy*
- *successful experiences*

Water reuse projects are implemented for many reasons. These reasons include the following:

- Opportunity—Historically, many golf courses receiving reclaimed water have been located in areas close to existing treatment plants where additional treatment is not required for irrigation, and where extensive reclaimed water transmission and distribution lines are unnecessary. Under such conditions, reuse is more opportunistic in nature rather than the result of a well–planned program to supplement or replace the use of potable water for irrigation.

- Need—With increasing water demands in areas with limited freshwater supplies, reclaimed water may be the only cost–effective way of supplementing water resources.

- Conservation—Using reclaimed water for irrigation can significantly reduce the freshwater demand.

Reclaimed water users are usually assured of receiving their allocation of water.

Use of reclaimed water may be the most economical option for providing irrigation water to a golf course.

The public supports reuse and through elected officials is becoming more active in promoting it.

- Reliability of Supply—Reclaimed water users are usually assured of receiving their allocation of water. Interruptions in the production of reclaimed water are usually very short term, and can be overcome by providing adequate storage facilities or by having an alternative source of water for emergency situations. The reclaimed water supply may be more reliable than the freshwater supply in times of water shortage.

- Well–Established Technology—Water reclamation for nonpotable purposes, including golf course irrigation, only requires conventional water and wastewater treatment technology that has been proven to be effective and is well–established in the U.S.

- Economics—Use of reclaimed water may be the most economical option for providing irrigation water to a golf course. The producer may realize cost savings due to lessened treatment and disposal costs and the sale of the reclaimed water. Reclaimed water is typically priced less than potable water, sometimes as an incentive to use the water, thus resulting in cost savings to the user. The community benefits by eliminating or delaying the costs associated with obtaining additional sources of freshwater.

- Pollution Abatement—Using reclaimed water that would otherwise be discharged into environmentally–sensitive surface waters eliminates a source of contamination in those waters. It is often less costly to produce reclaimed water suitable for golf course irrigation than to provide a high level of treatment, e.g., nutrient removal necessary for discharge into many surface waters.

- Public Policy—The public supports reuse and through elected officials is becoming more active in promoting it. In water–short regions of the country it is becoming increasingly more common for state, regional, and local regulatory agencies to mandate water reclamation and reuse.

- Successful Experiences—While research and pilot plant studies provide useful and necessary information, the ultimate test is a full–scale operation. There are literally hundreds of successful water reclamation and reuse operations in the U.S.[6]

Detailed up–to–date information on the number of golf courses irrigated with reclaimed water on a national level is not available, although there are well over 200 such projects in the U.S. As would be expected, irrigation with reclaimed water is extensively practiced in the arid west and southwestern states of the U.S. However, the advantages of water reclamation and reuse have been recognized in other parts of the country as well, where projects have been implemented to extend local water supplies or reduce or eliminate surface water discharges. In south Florida, for example, subtropical conditions and almost 55 inches/year (140 cm/yr) of rainfall suggest an abundance of water; but cultural practice and regional hydrogeology combine to result in frequent water shortages and restrictions on water use. As a result, water reuse, particularly for golf course irrigation, is widely practiced in Florida.

There are literally hundreds of successful water reclamation and reuse operations in the U.S.

Need for Regulations

While water reclamation and reuse regulations clearly are needed to assure public health protection, they also provide consistency of regulatory decisions and requirements for all projects. With regulations in place, project proponents can determine and evaluate economic and other effects of regulatory constraints and restrictions early in the planning process, which may be key factors in the decision–making process.

The public's acceptance and support of reuse projects is based, in part, on its confidence in the safety of reclaimed water. The public will support the use of reclaimed water for golf course irrigation if the state's health and other regulatory agencies provide safety assurances via appropriate water reclamation and reuse regulations. The absence of regulations may be interpreted by both regulators and potential users as a prohibition of golf course irrigation with reclaimed water.

The public will support the use of reclaimed water for golf course irrigation if the state's health and other regulatory agencies provide safety assurances via appropriate water reclamation and reuse regulations.

Regulations are preferable to guidelines.
- *guidelines are an indication of policy or conduct and are nonenforceable*
- *regulations are enforceable rules or standards established by legislative authority*

Therefore, the existence of regulations that allow the use of reclaimed water for irrigation with proper restrictions and controls can provide an impetus for reuse and may encourage consideration of reuse as a viable, cost–effective option.

Regulations are preferable to guidelines. Guidelines are an indication of policy or conduct, while regulations are rules or standards established by legislative authority. In a regulatory context, guidelines for water reclamation and reuse are advisory, voluntary, and nonenforceable. By contrast, regulations are mandatory, enforceable rules or standards. Application of guidelines strictly as recommended goals seldom occurs in practice, and most states with guidelines include them as part of reuse permit requirements.

Water Quality Considerations

The presence of toxic chemicals and pathogenic microorganisms in untreated wastewater creates the potential for adverse health effects where there is contact, inhalation, or ingestion of chemical or microbiological constituents of health concern. Thus, the acceptability of reclaimed water for golf course irrigation is dependent on the physical, chemical, and microbiological quality of the water. The effects of physical parameters, e.g., pH, color, temperature, and particulate matters, and chemical constituents, e.g., chlorides, sodium, heavy metals, and trace organics, on turf, other vegetation, soil and groundwater are well known, and recommended limits have been established for many constituents.[15,27,33,31]

The acceptability of reclaimed water for golf course irrigation is dependent on the physical, chemical, and microbiological quality of the water.

In contrast to the agronomic concerns associated with chemical constituents that may be present in wastewater, microbiological constituents present health considerations for the distribution and use of reclaimed water. Domestic sewage can contain a myriad of disease–causing organisms; hence, water reuse standards and guidelines are principally directed at public health protection.

Factors that affect the quality of reclaimed water include source water quality, wastewater treatment processes and treatment effectiveness, treatment reliability, and distribution system design and operation. Industrial source control programs can limit the input of chemical constituents that may present health, environmental, or irrigation concerns or that may adversely affect treatment processes and subsequent acceptability of the water for specific uses. Assurance of treatment reliability is an obvious, yet sometimes overlooked, quality control measure. Distribution system design and operation is important to reduce the possibility of cross–connections to potable water supplies and assure that the reclaimed water is not degraded prior to use and not subject to accidental or deliberate misuse. Open storage may result in water quality degradation by microorganisms, algae, or particulate matter, and may cause objectionable odor or color in the reclaimed water.

Factors that affect reclaimed water quality include:
  • source water
  • treatment processes
  • treatment reliability
  • distribution system

For golf course irrigation, considerations for water quality criteria include the following:

• Public Health Protection—Reclaimed water must be safe for the intended use. Most reclaimed water criteria are principally directed at public health protection, and many address only microbiological concerns.

• Use Requirements—There are specific physical and chemical water quality requirements not related to health considerations. The effect of individual constituents or parameters on turf or other vegetation, soil, and groundwater or other receiving water must be considered. State water reuse regulations generally address groundwater contamination concerns.

• Environmental Considerations—Natural flora and fauna in and around golf courses should not be adversely affected by reclaimed water. Runoff to surface water should be controlled.

• Aesthetics—Ideally, reclaimed water should not be different in appearance, i.e., clear, colorless, and odorless, than potable water or other sources of water used for irrigation. When used in water hazards, reclaimed water should not promote algal growth.

Golf course irrigation considerations include:
  • public health
    protection
  • use requirements
  • environmental
  • aesthetics
  • public perception
  • political realities

The water must be perceived as being safe and acceptable for golf course irrigation.

- Public and/or User Perception—The water must be perceived as being safe and acceptable for golf course irrigation. This may result in the imposition of conservative quality limits by regulatory agencies.

- Political Realities—Regulatory decisions are sometimes based on the political climate, public policy, personal beliefs or biases, and cost.

Water reuse standards and guidelines are principally directed at public health protection.

Key factors in the establishment of water reclamation and reuse criteria include health protection, public policy, past reuse experience, and economics. Adverse health consequences associated with the reuse of raw or improperly treated wastewater are well–documented.[9,14,19] As a result, water reuse standards and guidelines are principally directed at public health protection and are generally based on the control of pathogenic bacteria, protozoa, helminths, and viruses.

Treatments Reliability

Water reuse requires strict conformance to all applicable water quality limits.

Water reuse requires strict conformance to all applicable water quality limits, as there is potential for harm in the event that improperly treated reclaimed water is delivered to use areas. The need for reclamation facilities to reliably and consistently produce and distribute reclaimed water of adequate quality and quantity is essential and dictates that careful attention be given to reliability features during the design, construction, and operation of the facilities.

EPA Class I reliability as defined by the EPA is generally recommended for reclaimed water production. Class I reliability requires redundant treatment unit processes and equipment to prevent treatment upsets during power and equipment failures, flooding, peak loads, and maintenance shutdowns.[26] Other necessary design features include the following:

- duplicate power sources

- standby power for essential treatment plant elements

- emergency storage or disposal of inadequately–treated wastewater

- piping and pumping flexibility to permit rerouting of flows under emergency conditions

- provisions for uninterrupted chlorine feed, including standby chlorine supply, manifold systems to connect chlorine cylinders, chlorine weighing scales, and automatic devices for switching to full chlorine cylinders

- automatic alarms

The major concern which guides design, construction, and operation of a reclaimed water distribution system is the prevention of cross–connections.

Non–design reliability features include provisions for qualified personnel, an effective monitoring program, a quality assurance program, and an effective maintenance and process control program. An industrial pretreatment program, which is required in most states, and enforcement of sewer use ordinances to prevent illicit dumping of hazardous materials into the collection also contribute to treatment reliability.

Conveyance and Distribution Facilities

The distribution network includes pipelines and appurtenances, pumping stations, and storage facilities. The major concern which guides design, construction, and operation of a reclaimed water distribution system is the prevention of cross–connections. A cross–connection is a physical connection between a potable water system and any source containing nonpotable water through which potable water could be contaminated. Another major concern of regulatory agencies is improper use or inadvertent use of reclaimed water.

Typical regulatory controls to prevent cross–connections and intentional or unintentional misuse of reclaimed water include the following: identification of pipes and appurtenances; horizontal and vertical separation of potable and reclaimed water lines; prevention of ability to tie into reclaimed water lines; backflow protection devices on potable water lines; and pipeline design and construction criteria.

Controls to prevent cross–connections include:
- *proper identification*
- *horizontal and vertical separation*
- *prevention of easy connection to reclaimed water source*
- *backflow protection*
- *pipeline design and construction*

All reclaimed water pipelines and appurtenances should be clearly identified.

All reclaimed water pipelines and appurtenances, e.g., pumps, outlets, and valve boxes, should be clearly identified. Some states specify signs, tags, color–coding, marking tape, and/or stenciled wording. For example, the color purple is used in California to identify reclaimed water lines and appurtenances, and other states are beginning to use the same color. Valve covers on reclaimed water distribution systems should not be interchangeable with potable valve covers. Where fire hydrants are part of a reclaimed water system, they should be painted or marked to identify them as such and the stem should require a special wrench for opening.

Some states require specific horizontal and vertical separation between reclaimed and potable pipelines.

To assure separation between reclaimed and potable pipelines, some states require a 10–foot (3–m) horizontal separation and a 1–foot (0.3–m) vertical separation where the lines are parallel to each other. Where these distances cannot be maintained, special authorization may be required, although a minimum lateral distance of 4 feet (1.2 m) is generally mandatory[28]. Where reclaimed and potable water lines cross, the reclaimed water line is generally required to be at least 1 foot (0.3 m) below the potable line. Often, special construction techniques can be utilized to reduce this distance. Reclaimed water lines are usually required to be at least 3 feet (0.9 m) below the ground surface.

Several states do not permit hose bibbs on reclaimed water systems because of the potential for misuse.

Several states do not permit hose bibbs on reclaimed water systems because of the potential for misuse. Florida regulations do allow hose bibbs in below–ground lockable vaults or where a special tool is required to obtain access to the reclaimed water[23]. Special quick–coupling valves may be required for onsite irrigation connections in some states.

Backflow prevention is needed to protect potable water lines at use areas that receive both potable and reclaimed water.

Except in special cases, some form of backflow prevention is needed to protect potable water lines at use areas that receive both potable and reclaimed water. Most states require installation of a backflow prevention device on the potable water service line to prevent potential backflow from the reclaimed water system into the potable water system if the two systems are inadvertently or intentionally interconnected. The American Water Works Association recommends the use of a reduced pressure principle backflow prevention device where reclaimed water systems are present[1].

Water Reclamation and Reuse Regulations/Guidelines

There are no federal regulations governing water reclamation and reuse in the U.S.; hence, the regulatory burden rests with the individual states. This has resulted in widely differing standards, and many states do not have any standards or guidelines for water reuse. The absence of standards can inhibit the development and implementation of potential projects. The absence of standards may be viewed by some as prohibiting reuse or allowing regulatory agencies to make ad hoc decisions on specific projects. Without discrete regulations, requirements could change significantly in time, thus presenting a subjective "moving target."

There are no federal regulations governing water reclamation and reuse in the U.S.; hence, the regulatory burden rests with the individual states.

Several water–short states have comprehensive regulations and prescribe requirements according to the end use of the water. Some states have regulations or guidelines directed at land treatment or land application of wastewater rather than the beneficial use of reclaimed water, while several states have no regulations or guidance documents. In 1992, 18 states had some form of water reuse regulations, 18 states had guidelines, and 14 states had neither regulations or guidelines[28]. Table 1 summarizes the extent of state regulations and guidelines in effect as of March 1992.

The absence of standards can inhibit the development and implementation of potential projects.

Prior to release of the recently-published *Guidelines for Water Reuse*[28], a document jointly sponsored by EPA and the U.S. Agency for International Development, states received little guidance from federal agencies for criteria development. As a consequence, there is no consistency among the various states' water reuse criteria. Criteria tend to become more complete and, in some cases, more conservative as the number of reuse projects increases in a state and more attention is given to providing suitable controls to assure that public health is not compromised.

In 1992, 18 states had some form of water reuse regulations, 18 states had guidelines, and 14 states had neither regulations or guidelines.

Table 1. Summary of State Reuse Regulations and Guidelines Source: U.S. Environmental Protection Agency, 1992

State	Regulations	Guidelines	No Regulations or Guidelines	Unrestricted Urban Reuse	Restricted Urban Reuse	Agricultural Reuse Food Crops	Agricultural Reuse Non-Food Crops	Unrestricted Recreational Reuse	Restricted Recreational Reuse	Environmental Reuse	Industrial Reuse
Alabama		●									
Alaska			●								
Arizona	●	●		●	●	●	●	●	●	●	
Arkansas		●		●	●	●	●				
California	●	●		●	●	●	●	●	●		
Colorado			●	●	●	●	●	●	●	●	
Connecticut							●				
Delaware		●				●	●				
Florida	●			●	●	●	●				
Georgia		●		●	●		●				
Hawaii		●(1)		●	●	●	●		●		●
Idaho	●(1)						●				
Illinois	●(1)						●				
Indiana	●						●				
Iowa			●								
Kansas		●		●	●	●	●				
Kentucky			●								
Louisiana			●								
Maine		●					●				
Maryland					●		●				
Massachusetts			●								
Michigan	●					●	●				
Minnesota			●								
Mississippi			●		●						
Missouri	●						●				

(1) Draft or Proposed

Table 1. Summary of State Reuse Regulations and Guidelines (Continued)

State	Regulations	Guidelines	No Regulations or Guidelines	Unrestricted Urban Reuse	Restricted Urban Reuse	Agricultural Reuse Food Crops	Agricultural Reuse Non-Food Crops	Unrestricted Recreational Reuse	Restricted Recreational Reuse	Environmental Reuse	Industrial Reuse
Montana		●				●	●				
Nebraska		●			●		●				
Nevada		●(1)		●	●	●	●	●	●		●
N. Hampshire			●								
New Jersey	●						●				
New Mexico		●		●	●	●	●				
New York		●					●				
N. Carolina	●				●		●				
N. Dakota		●					●				
Ohio		●	●								
Oklahoma	●				●	●	●	●	●		●
Oregon		●		●	●	●	●				
Pennsylvania			●								
Rhode Island		●	●								
S. Carolina		●		●	●		●				
S. Dakota	●	●		●	●		●			●	
Tennessee	●			●	●	●	●				
Texas	●			●	●	●	●	●	●		●
Utah	●			●	●		●				●
Vermont	●						●				
Virginia			●								
Washington		●		●	●	●	●				
West Virginia	●					●	●				
Wisconsin	●						●				
Wyoming	●			●	●	●	●				

(1) Draft or Proposed

Arizona, California, Florida and Texas have well–developed water reuse programs.

Regulations in states having well–developed water reuse programs, i.e., Arizona, California, Florida, and Texas, are discussed below, as are regulations for the State of North Carolina, where there is little reuse activity at the present time. EPA's recently–published water reuse guidelines are also presented. These examples are intended to provide an overview of various state standards, with particular emphasis on treatment and quality requirements. A more complete summary of all state regulations and guidelines is provided in the *Guidelines for Water Reuse*[28].

Arizona

Arizona is the only state having water reuse criteria that include limits for viruses and parasites.

Arizona is the only state having water reuse criteria that include limits for viruses and parasites[20]. The present regulations do not require specific wastewater treatment unit processes. The Arizona regulations are currently under revision and will include both treatment process requirements and water quality requirements, and the virus and parasite monitoring requirements are being dropped from the regulations[17]. The latest draft revisions to the Arizona regulations are based on, and are similar to, recommended criteria contained in EPA's *Guidelines for Water Reuse*[28].

California

The State of California has a long history of reuse and first developed reuse regulations in 1918, which have been modified and expanded through the years. The state's current Wastewater Reclamation Criteria[21], which are in the process of being revised, were adopted in 1978 and have served as the basis for reuse standards in other states and countries. The reclamation criteria include water quality standards, treatment process requirements, operational requirements, and treatment reliability requirements. The treatment and quality criteria are shown in Table 2.

The reclamation criteria allow golf course irrigation with reclaimed water that is oxidized and disinfected such that the median number of total coliform organisms does not exceed 23/100 mL, as determined from the last 7 days for which analyses have been completed. Daily coliform sampling is required. While reclaimed water meeting these requirements is not assuredly pathogen–free, it is deemed to be safe for the intended use if appropriate use area controls, e.g., off–hours irrigation, are instituted. These requirements are appropriate for golf courses in rural settings, where windblown spray does not reach residential property or other areas frequented by the public.

Irrigation of golf courses in urban areas may necessitate more stringent wastewater treatment and quality standards.

Many golf courses constructed in recent years are located adjacent to urban areas or have residential lots abutting the courses. In such cases, occasional direct or indirect human contact with reclaimed water is likely, and California imposes more stringent wastewater treatment and quality requirements than those specified in the Wastewater Reclamation Criteria for golf course irrigation. The wastewater must be oxidized, coagulated, filtered, and disinfected such that the 7–day median number of total coliform organisms does not exceed 2.2/100 mL.

The coliform levels in Table 2 are not definitive threshold levels justified by rigorous documentation and evaluation of illness rates. At the time the regulations were developed, the California Department of Health Services (DOHS) concluded that epidemiological studies of the exposed population at water reuse sites would be of limited value, and that it was not possible to ascribe numerical risk estimates to reclaimed water with any degree of confidence. Thus, the reclamation criteria were based on the capability of well–designed and well–operated wastewater treatment plants to consistently attain specific effluent quality limits, experience at existing wastewater disposal and reuse operations, evaluation of pertinent research studies and health–related data, and the desire not to allow unreasonable risks due to the use of reclaimed water. The intent of the regulations is to "establish acceptable levels of constituents of reclaimed water and to prescribe means for assurance of reliability in the production of reclaimed water in order to ensure that the use of reclaimed water for the specified purposes does not impose undue risks to health[21]."

California regulations establish acceptable levels of constituents within reclaimed water in order to ensure that its use for the specified purpose does not impose undue risks to health.

Table 2. California Treatment and Quality Criteria for Reuse

Type of Use	Total Coliform Limits	Treatment Required
Fodder, Fiber, and Seed Crops Surface Irrigation of Orchards and Vineyard	—	Primary
Pasture for Milking Animals Landscape Impoundments Landscape Irrigation (Golf Courses, Cemeteries, etc.)	23/100 mL	Oxidation Disinfection
Surface Irrigation of Food Crops Restricted Recreational Impoundments	2.2/100 mL	Oxidation Disinfection
Spray Irrigation of Food Crops Landscape Irrigation (Parks, Playgrounds, etc.) Nonrestricted Recreational Impoundments	2.2/100 mL	Oxidation Coagulation Clarification Filtration[a] Disinfection
Groundwater Recharge	Case-by-Case Evaluation	Case-by-Case Evaluation

[a]The turbidity of filtered effluent cannot exceed an average of 2 turbidity units during any 24-hour period.

Source: State of California, 1978

As indicated in Table 2, the required degree of treatment and microbiological quality increase as the likelihood of human exposure to the reclaimed water increases. If intimate direct contact with the reclaimed water is expected, such as swimming, or indirect contact is likely, such as consuming produce spray–irrigated with reclaimed water, the regulations specify treatment and water quality requirements intended to produce an effluent that is essentially free of measurable levels of pathogens, including viruses. A fundamental decision was made that the standard to be applied was to be the absence of measurable levels of viruses, based on the assumptions that very low numbers of viruses can initiate infection and wastewater treatment processes assuredly controlling viruses would produce reclaimed water free from any human pathogen and thus be safe for the intended uses.

The required degree of treatment and microbiological quality increase as the likelihood of human exposure to the reclaimed water increases.

Selection of the treatment train specified in the Wastewater Reclamation Criteria to produce an essentially pathogen–free effluent, i.e., oxidation, chemical coagulation, clarification, filtration, and disinfection to a total coliform level not exceeding 2.2/100 mL, was predicated on studies conducted several years ago to determine the virus removal capability of tertiary wastewater treatment processes. More recent studies[8,18] indicated that equivalent virus removal can be achieved by direct filtration of high quality secondary effluent, using low coagulant and/or polymer dosages. This abbreviated treatment chain, in conjunction with specific design and operational controls, has been judged to be equivalent to the full treatment chain specified in the regulations[3]. The required design and operational controls for direct filtration facilities include the following:

Wastewater treatment processes may include
- *oxidation*
- *chemical coagulation*
- *clarification*
- *filtration*
- *disinfection*

- coagulant addition unless secondary effluent turbidity is less than 5 NTU

- maximum filtration rate of 5 gpm/ft^2(12 m/h)

- average filter effluent turbidity of 2 NTU or less

- high energy rapid mix of chlorine

- theoretical chlorine contact time of at least 2 hours with an actual modal contact time of at least 90 minutes.

- minimum chlorine residual of 5 mg/L after the required contact time

- chlorine contact chamber length to width or depth ratio of at least 40:1

- 7–day median number of total coliform organisms in the effluent of 2.2/100 mL or less, not to exceed 23/100 mL in any sample

Reliability require-
ments address
  • standby power
    supplies
  • alarm systems
  • multiple or standby
    treatment units
  • emergency storage
    or disposal of
    inadequately
    treated wastewater
  • prevention of
    treatment bypassing
  • monitoring devices
  • design flexibility

The Wastewater Reclamation Criteria also include requirements for treatment reliability. The reliability requirements address standby power supplies, alarm systems, multiple or standby treatment process units, emergency storage or disposal of inadequately treated wastewater, elimination of treatment process bypassing, monitoring devices and automatic controllers, and flexibility of design.

While the reclamation criteria do not specifically address reclaimed water use area controls, DOHS has established guidelines that describe safety precautions and operational procedures, such as cross–connection control provisions, confinement of reclaimed water at use areas, color–coded reclaimed water lines and appurtenances, separation and construction criteria for potable and reclaimed water pipelines, key–operated valves and outlets, fencing, signs, control of windblown spray, and provisions for worker protection[2]. Individual water reclamation requirements imposed by the Regional Water Quality Control Boards normally include specific use restrictions based on the DOHS use area guidelines.

Florida

The primary driving
force behind imple-
mentation of reuse
projects in Florida
was effluent disposal.

Until recently, the primary driving force behind implementation of reuse projects in Florida was effluent disposal[32]. Regulations developed in the early 1980s for reuse and land application were contained in a document entitled *Land Application of Domestic Wastewater Effluent in Florida*[10]. Irrigation of public access areas, which includes golf courses, and irrigation of edible crops were allowed but requirements for such activities were incomplete. Rule 17–610, Florida Administrative Code.

Reuse of Reclaimed Water and Land Application, was adopted in 1989 and revised in 1990. The treatment and quality criteria are shown in Table 3.

In addition to the wastewater treatment and water quality requirements presented in Table 3, Florida's reuse rule includes design and use area requirements such as the following:

- minimum system size of 0.1 mgd (380 m^3/d) for any public access irrigation system, e.g., golf courses, and 0.5 mgd (1,900 m^3/d) for any residential lawn or edible crop irrigation

- operating protocol that includes continuous turbidity and chlorine monitoring at water reclamation plants.

- 24–hour per day staffing at water reclamation plants

- reject storage capacity in a lined storage facility of at least 1 day to hold unacceptable quality product water for return for additional treatment

- minimum system storage of at least 3 days if no backup system is provided

- prohibition of cross–connections with potable systems. Double check valve backflow prevention devices are acceptable on potable water lines entering property served by reclaimed water systems

- ordinances or user agreements to document controls on individual users

- use area controls, including groundwater monitoring, surface runoff control, public notification, and setback distances

Reuse of reclaimed water from domestic wastewater treatment facilities is required within the designated critical water supply problem areas unless such reuse is not economically, environmentally, or technically feasible.

The Florida Water Policy[22] establishes a mandatory reuse program. The policy requires that the state's water management districts identify "critical water supply problem areas" that currently exist or are anticipated during the next 20 years. Reuse of reclaimed water from domestic wastewater treatment facilities is required within the designated critical water supply problem areas unless such reuse is not economically, environmentally, or technically feasible. The water management districts can

Table 3. Florida Treatment and Quality Criteria for Reuse

Type of Use	Allowable Limits	Treatment Required
Restricted Public Access Areas[a]	200 Fecal Coli/100 mL 20 mg/L TSS	Secondary 20 mg/L BOD Disinfection
Public Access Areas[b] Food Crop I Toilet Flushing[d] Fire Protection Aesthetic Purposes Dust Control	No Detectable Fecal Coli/100 mL rrigation[c] 5 mg/L TSS 20 mg/L BOD	Secondary Disinfection Filtration
Rapid Rate Land Application	200 Fecal Coli/100 mL 20 mg/L TSS 20 mg/L BOD 12 mg/L Total N	Secondary Disinfection

[a]Sod farms, forests, fodder crops, pasture land, or similar areas.
[b]Residential lawns, golf courses, cemeteries, parks, landscaped areas, highway medians, or similar areas.
[c]Only allowed if crops peeled, skinned, cooked or thermally processed before consumption.
[d]Not allowed where residents have access to plumbing system.

Source: State of Florida, 1990

require reuse of reclaimed water outside of designated critical water supply areas if the reclaimed water is readily available and the water management district has adopted rules for reuse in those areas.

The Florida Department of Environmental Protection has developed *Guidelines for Preparation of Reuse Feasibility Studies for Applicants Having Responsibility for Wastewater Management*[12]. The document provides comprehensive requirements for the preparation of reuse feasibility studies. The requirements include evaluation of alternatives, including at least a "no action" alternative and a public access/urban reuse system alternative. The feasibility evaluation must include economic considerations, reuse benefits, and technical feasibility. Under certain conditions, a municipality, governmental entity, or utility that has developed and is implementing a reuse master plan can have the master plan accepted in lieu of a reuse feasibility study report.

North Carolina

North Carolina provides a good example of a state that is just beginning to recognize water reclamation and reuse as a viable alternative to surface water discharge and as a valuable water resource in the future. Because the use of reclaimed water is a relatively new concept in the state, water reuse regulations are not yet well–developed or comprehensive in scope.

North Carolina provides a good example of a state that is just beginning to recognize water reclamation and reuse, both as a viable alternative to surface water discharge and as a valuable water resource in the future.

In 1976, the State of North Carolina adopted regulations governing "waste not discharged to surface waters". The current regulations, North Carolina Administrative Code Section T15A: 02H .0200, were last revised in 1987[24]. These regulations principally address animal waste management systems, although they do contain minimal requirements for land application of domestic wastewater on golf courses and other public access areas.

The following requirements are applicable for the irrigation of golf courses and other public areas with reclaimed water:

- aerated flow equalization facilities with a capacity of at least 25 percent of the daily system design flow

- all essential treatment and disposal units shall be provided in duplicate

- the treatment process shall produce an effluent with a monthly average total suspended solids (TSS) of less than 5 mg/L, a daily maximum TSS of less than 10 mg/L, and a maximum fecal coliform level of less than 1/100 mL, prior to discharge to a five–day detention pond

- there shall be no public access to the five–day detention pond

- the size of the irrigation pond that follows the five–day holding pond shall be justified using a mass water balance for worse case conditions on record

- an automatically–activated standby power source or other means to prevent improperly–treated wastewater from entering the five–day detention pond shall be provided

- requirements for the lining of five–day detention ponds and irrigation ponds shall be site–specific

- in the design of the sprinkler system, the piping shall be a separate system, with no cross–connections to a potable water supply (includes no spigots on the distribution system)

- the rate of application shall be site–specific but not exceed 1.75 inches/week

- the time of spraying shall occur between 11:00 p.m. and three hours prior to the daily opening of the course

- there shall be a 100–foot vegetative buffer zone between the edge of spray influence and the nearest dwelling

- signs shall be posted at the pro shop stating that

the course is irrigated with treated wastewater

- there shall be a certified operator of a class equivalent to the class plant on call 24 hours per day

Other requirements in the regulations address plans and
specifications, permitting requirements, chemical analyses of raw wastewater, protection from a 100–year flood,
and monitoring well requirements. While the regulations
do not specify level of treatment or unit processes, it is
likely that secondary treatment followed by filtration and
a high level of disinfection will be necessary to consistently achieve the TSS and fecal coliform requirements
for reclaimed water used for golf course and other public
access area irrigation. The absence of requirements for
other uses of reclaimed water is not meant to imply that
other uses are prohibited; such uses will be evaluated on
a case–by–case basis.

In North Carolina, secondary treatment followed by filtration and a high level of disinfection will be necessary to consis- tently achieve the TSS and fecal coliform requirements for reclaimed water used for golf courses.

In early 1992, the Division of Environmental Management (DEM) of the North Carolina Department of Environmental, Health, and Natural Resources (DEHNR)
developed proposed amendments to the nondischarge
rules in 1992. DEM recommended that revisions presented in the Environmental Management Commission's
*Report of Proceedings for the Proposed Revisions to the
Nondischarge Rules* be adopted[16]. The proposed revisions to requirements that affect water reclamation and
reuse are presented below.

Existing Regulations	Proposed Revisions
<1 fecal coli/100 mL	<5 fecal coli/100 mL
Max. application rate = 1.75 inches/week	Application rate site specific
Irrigate between 11:00 p.m. and 3 hours prior to opening golf course	Not stated
100–foot buffer zone between edge of spray area and dwellings	50–foot buffer zone between edge of spray and dwellings

Texas

Reclaimed water regulations in Texas include golf course in the definition of a "restricted landscape area."

Reclaimed water regulations in Texas include golf course in the definition of a "restricted landscape area," that is . . . "Land which has had its plant cover modified and access to which may be controlled in some manner. Access may be controlled by either legal means (e.g., state or city ordinances) or controlled by some type of physical barrier (e.g., fence or wall). Examples of such areas are: golf courses; cemeteries; roadway right–of–ways; median dividers."[25].

The regulations include water quality criteria and design and operational requirements. The water quality criteria are summarized in Table 4. In contrast to other states that have extensive reuse and comprehensive regulations, the Texas regulations do not specify wastewater treatment processes and allow a relatively low quality effluent to be used for golf course irrigation. Irrigation is acceptable provided the biochemical oxygen demand (BOD) of the reclaimed water does not exceed 20 mg/L and the fecal coliform level does not exceed 800 colony-forming units/ 100 mL. Both of these limits are based on a 30–day average. When stabilization ponds are used as a treatment process, a 30–day average BOD not exceeding 30 mg/ L is acceptable.

Texas regulations do not specify wastewater treatment processes and allow a relatively low quality effluent to be used for golf course irrigation.

If reclaimed water is stored for 24 hours or longer prior to irrigation, the water must be disinfected as needed to meet the fecal coliform limits. In other major reuse states, water quality requirements apply to reclaimed water as it leaves the treatment plant and not at the point of use. Those states require a considerably higher microbiological quality for golf course irrigation with reclaimed water and, on the premise that most pathogenic agents potentially present in untreated sewage have been destroyed, consider that any regrowth would be confined to nonpathogenic microorganisms and thus not represent a health threat. Any contamination occurring in storage ponds from birds, runoff, etc., would be no different than would occur in fresh water that is used for irrigation and exposed to the same environmental conditions.

If reclaimed water is stored for 24 hours or longer prior to irrigation, the water must be disinfected as needed to meet the fecal coliform limits.

Use area controls affecting golf course irrigation, many of which are similar to those imposed by other states, include the following:

- the irrigation site must be maintained with a vegetative cover or be under cultivation during times when reclaimed water is applied

- the irrigation practice must be designed to prevent incidental ponding or standing water

- golf courses can only be irrigated when not in use

- if irrigation water is stored prior to application, provision must be made to provide additional disinfection to meet the specified criteria for the designated area

- reclaimed water spray must not reach any privately–owned premises outside the designated irrigation area or public drinking fountains

- irrigation is prohibited when the ground is saturated or frozen

- tailwater must be controlled to preclude discharge of reclaimed water from the irrigation site

- distribution systems must be designed to prevent operation by unauthorized personnel

Contracts are required between reclaimed water providers and users to identify their respective responsibilities and liabilities. For example, reclaimed water can be transferred from a provider to a user on a demand basis only, and a user may refuse delivery of reclaimed water at any time. A detailed water balance is required to determine the appropriate reclaimed water application rate.

EPA Guidelines

The EPA recently published a document entitled *Guidelines for Water Reuse*[28], which was prepared by Camp Dresser & McKee Inc. with significant input from a

The EPA recently published guidelines that address
- *important aspects of wastewater reuse*
- *recommended treatment processes*
- *treatment reliability provisions*
- *water quality limits*
- *monitoring frequencies*
- *setback distances*
- *other controls for water reuse applications*

Table 4 Texas Reclaimed Water Quality Requirements

Type of Use	Allowable Limits
Irrigation	
Fodder, Fiber, & Seed Crops	30 mg/L BOD
Pasture for Milking Animals	20 mg/L BOD[a] 800 Fecal Coli/100 mL
Food Crops[b]	10 mg/L BOD[a] 3 NTU 75 Fecal Coli/100 mL
Restricted Landscape Areas	20 mg/L BOD[a] 800 Fecal Coli/100 mL
Unrestricted Landscape Areas	5 mg/L BOD 3 NTU 75 Fecal Coli/100 mL
Landscape Impoundments Restricted Landscape Impoundments Ornamental Fountains	10 mg/L BOD 3 NTU 75 Fecal Coli/100 mL
Commercial and Industrial Uses	20 mg/L BOD[a] 200 Fecal Coli/100 mL
Toilet Flush Water	5 mg/L BOD 75 Fecal Coli/100 mL

[a]30 mg/L BOD for stabilization pond systems.
[b]Spray irrigation of food crops eaten raw is prohibited.

Source: State of Texas, 1990.

technical advisory committee of more than 50 nationally and internationally recognized experts in public health, water resources, wastewater engineering, regulatory activities, and other areas of water reclamation and reuse. The guidelines address all important aspects of water reuse, including recommended wastewater treatment processes, treatment reliability provisions, reclaimed water quality limits, monitoring frequencies, setback distances, and other controls for various water reuse applications. The suggested treatment and water quality guidelines are presented in matrix format in Table 5. They are intended to be somewhat conservative and provide reasonable guidance for states that have not developed their own criteria or guidelines.

The EPA guidelines are somewhat conservative and provide reasonable guidance for states that have not developed their own guidelines.

Both wastewater treatment and reclaimed water quality limits are recommended for the following reasons: water quality criteria involving surrogate parameters do not adequately characterize reclaimed water quality; a combination of treatment and quality requirements known to produce reclaimed water of acceptable quality obviate the need to monitor the finished water for certain constituents; expensive, time–consuming, and in some cases, questionable monitoring for pathogenic microorganisms is eliminated without compromising health protection; and treatment reliability is enhanced[15].

The guidelines include limits for fecal coliform organisms but do not include parasite or virus limits. Parasites have not been shown to be a problem at reuse operations in the U.S. While viruses are a concern in reclaimed water, virus limits are not recommended in the guidelines for the following reasons: a significant body of information exists indicating that viruses are inactivated or removed to low or immeasurable levels via appropriate wastewater treatment[5,8,18]; the identification and enumeration of viruses in wastewater are hampered by relatively low virus recovery rates; there are a limited number of facilities having the personnel and equipment necessary to perform the analyses; the laboratory analyses can take as long as four weeks to complete; there is no consensus among public health experts regarding the health significance of low levels of viruses in reclaimed water[7]; and

For golf courses, the EPA guidelines include:
- *limits for fecal coliform organisms*
- *recommendations that wastewater receive secondary treatment, filtration, and disinfection*
- *setback distances between wastewater irrigated areas and potable water sources*
- *maintenance of minimum chlorine residual*

Table 5 EPA Guidelines for Water Reuse Source: U.S. Environmental Protection Agency, 1992

Types of Reuse	Treatment	Reclaimed Water Quality[2]	Reclaimed Water Monitoring	Setback Distances[3]	Comments
Urban Reuse All types of landscape irrigation, (e.g. golf courses, parks, cemeteries) - also vehicle washing, toilet flushing, use in fire protection systems and commercial air conditioners, and other uses with similar access or exposure to the water	•Secondary[4] •Filtration[5] •Disinfection[6]	•pH = 6 - 9 •≤ 10 mg/L BOD[7] •≤ 2 NTU[8] •No detectable fecal coli/100 mL[9,10] •1 mg/L Cl₂ residual (min.)[11]	•pH - weekly •BOD - weekly •Turbidity - continuous •Coliform - daily •Cl₂ residual - continuous	•50 ft. (15 m) to potable water supply wells	• At controlled-access irrigation sites where design and operational measures significantly reduce the potential of public contact with reclaimed water, a lower lever of treatment, e.g., secondary treatment and disinfection to achieve ≤ 14 fecal coli/100 mL, may be appropriate. • Chemical (coagulant and/or polymer) addition prior to filtration may be necessary to meet water quality recommendations. • The reclaimed water should not contain measurable levels of pathogens.[12] • Reclaimed water should be clear, odorless, and contain no substances that are toxic upon ingestion. • A higher chlorine residual and/or a longer contact time may be necessary to assure that viruses and parasites are inactivated or destroyed. • A chlorine residual of 0.5 mg/L or greater in the distribution system is recommended to reduce odors, slime, and bacterial regrowth.
Restricted Access Area Irrigation Sod farms, silviculture sites, and other areas where public access is prohibited, restricted or infrequent	•Secondary[4] •Disinfection[5]	•pH = 6 - 9 •≤ 30 mg/L BOD[7] •≤ 30 mg/L SS •≤ 200 fecal coli/100 mL[9,13,14] •1 mg/L Cl₂ residual (min.)[11]	•pH - weekly •BOD - weekly •SS - daily •Coliform - daily •Cl₂ residual - continuous	•300 ft. (90 m) to potable water supply wells •100 ft. (30 m) to areas accessible to the public (if spray irrigation)	• If spray irrigation, SS less than 30 mg/L may be necessary to avoid clogging of sprinkler heads.
Agricultural Reuse - Food Crops Not Commercially Processed[15] Surface or spray irrigation of any food crop, including crops eaten raw	•Secondary[4] •Filtration[5] •Disinfection[6]	•pH = 6-9 •≤ 10 mg/L BOD[7] •≤ 2 NTU[8] •No detectable fecal coli/100 mL[9,10] •1 mg/L Cl₂ residual (min.)	•pH - weekly •BOD - weekly •Turbidity - continuous •Coliform - daily •Cl₂ residual - continuous	•50 ft. (15 m) to potable water supply wells	• Chemical (coagulant and/or polymer) addition prior to filtration may be necessary to meet water quality recommendations. • The reclaimed water should not contain measurable levels of pathogens.[12] • A higher chlorine residual and/or a longer contact time may be necessary to assure that viruses and parasites are inactivated or destroyed. • High nutrient levels may adversely affect some crops during certain growth stages.

Table 5 EPA Guidelines for Water Reuse (Continued)

Types of Reuse	Treatment	Reclaimed Water Quality[2]	Reclaimed Water Monitoring	Setback Distances[3]	Comments
Agricultural Reuse - Food Crops Commercially Processed,[15] Surface Irrigation of Orchards and Vineyards	•Secondary[4] •Disinfection[6]	•pH = 6 - 9 •≤ 30 mg/L BOD[7] •≤ 30 mg/L SS •≤ 200 fecal coli/100 mL[9,13,14] •1 mg/L Cl₂ residual (min.)[11]	•pH - weekly •BOD - weekly •SS - daily •Coliform - daily •Cl₂ residual - continuous	•300 ft. (90 m) to potable water supply wells •100 ft. (30 m) to areas accessible to the public (if spray irrigation)	• If spray irrigation, SS less than 30 mg/L may be necessary to avoid clogging of sprinkler heads. • High nutrient levels may adversely affect some crops during certain growth stages.
Agricultural Reuse - Non-Food Crops Pasture for milking animals; fodder, fiber and seed crops	•Secondary[4] •Disinfection[5]	•pH = 6 - 9 •≤ 30 mg/L BOD[7] •≤ 30 mg/L SS •≤ 200 fecal coli/100 mL[9,13,14] •1 mg/L Cl₂ residual (min.)[11]	•pH - weekly •BOD - weekly •SS - daily •Coliform - daily •Cl₂ residual - continuous	•300 ft. (90 m) to potable water supply wells •100 ft. (30 m) to areas accessible to the public (if spray irrigation)	• If spray irrigation, SS less than 30 mg/L may be necessary to avoid clogging of sprinkler heads. • High nutrient levels may adversely affect some crops during certain growth stages. • Milking animals should be prohibited from grazing for 15 days after irrigation ceases. A higher level of disinfection, e.g., to achieve ≤ 14 fecal coli/100 mL, should be provided if this waiting period is not adhered to.
Recreational Impoundments Incidental contact (e.g., fishing and boating) and full body contact with reclaimed water allowed	•Secondary[4] •Filtration[5] •Disinfection[6]	•pH = 6-9 •≤ 10 mg/L BOD[7] •≤ 2 NTU[8] •No detectable fecal coli/100 mL[9,10] •1 mg/L Cl₂ residual (min.)	•pH - weekly •BOD - weekly •Turbidity - continuous •Coliform - daily •Cl₂ residual - continuous	•500 ft. (150 m) to potable water supply wells (minimum) if bottom not sealed.	• Dechlorination may be necessary to protect aquatic species of flora and fauna. • Reclaimed water should be non-irritating to skin and eyes. • Reclaimed water should be clear, odorless, and contain no substances that are toxic upon ingestion. • Nutrient removal may be necessary to avoid algae growth in impoundments. • Chemical (coagulant and/or polymer) addition prior to filtration may be necessary to meet water quality recommendations. • The reclaimed water should not contain measurable levels of pathogens.[12] • A higher chlorine residual and/or a longer contact time may be necessary to assure that viruses and parasites are inactivated or destroyed. • Fish caught in impoundments can be consumed.

Table 5 EPA Guidelines for Water Reuse (Continued)

Types of Reuse	Treatment	Reclaimed Water Quality[2]	Reclaimed Water Monitoring	Setback Distances[3]	Comments
Landscape Impoundments Aesthetic impoundment where public contact with reclaimed water is not allowed.	•Secondary[4] •Disinfection[6]	•≤ 30 mg/L BOD[7] •≤ 30 mg/L SS •≤ 200 fecal coli/100 mL [9,13,14] •1 mg/L Cl$_2$ residual (min.)[11]	•pH - weekly •SS - daily •Coliform - daily •Cl$_2$ residual - continuous	•500 ft. (15 m) to potable water supply wells (minimum) if bottom not sealed	• Nutrient removal may be necessary to avoid algae growth in impoundments. • Dechlorination may be necessary to protect aquatic species of flora and fauna.
Construction Uses Soil compaction, dust control, washing aggregate, making concrete.	•Secondary[4] •Disinfection[5]	•≤ 30 mg/L BOD[7] •≤ 30 mg/L SS •≤ 200 fecal coli/100 mL [9,13,14] •1 mg/L Cl$_2$ residual (min.)	•BOD - weekly •SS - daily •Coliform - daily •Cl$_2$ residual-continuous		• Worker contact with reclaimed water should be minimized. • A higher level of disinfection, e.g., to achieve ≤ 14 fecal coli/100 mL, should be provided where frequent worker contact with reclaimed water is likely.
Industrial Reuse Once-through cooling	•Secondary[4]	•pH = 6-9 •≤ 30 mg/L BOD[7] •≤ 30 mg/L SS •≤ 200 fecal coli/100 mL [9,13,14] •1 mg/L Cl$_2$ residual (min.)	•pH - daily •BOD - weekly •SS - weekly •Coliform - daily •Cl$_2$ residual - continuous	•300 ft. (90 m) to areas accessible to the public.	• Windblown spray should not reach areas accessible to users or the public.
Recirculating cooling towers	•Secondary[4] • Disinfection[5] (chemical coagulation and filtration[5] may be needed)	•Variable, depends on recirculation ratio		•300 ft. (90 m) to areas accessible to the public. May be reduced if high level of disinfection is provided.	• Windblown spray should not reach areas accessible to users or the public. • Additional treatment by user is usually provided to prevent scaling, corrosion, biological growths, fouling and foaming.
Other Industrial Uses	← Depends on site specific use →	← Depends on site specific use →			

Table 5 EPA Guidelines for Water Reuse (Continued)

Types of Reuse	Treatment	Reclaimed Water Quality[2]	Reclaimed Water Monitoring	Setback Distances[3]	Comments
Environmental Reuse Wetlands, marshes, wildlife habitat, stream augmentation	•Variable •Secondary[4] • Disinfection[5] (min)	Variable, but not to exceed: •≤ 30 mg/L BOD[7] •≤ 30 mg/L SS •≤ 200 fecal coil/100 mL [9,13,14]	•BOD - weekly •SS - weekly •Coliform - daily •Cl₁ residual - continuous		• Dechlorination may be necessary to protect aquatic species of flora and fauna. • Possible effects on groundwater should be evaluated. • Receiving water quality requirements may necessitate additional treatment. • The temperature of the reclaimed water should not adversely affect ecosystem.
Groundwater Recharge By spreading or injection into non-potable aquifers.	•Site specific and use dependent •Primary (min.) for spreading •Secondary[4] (min.) for injection	•Site specific and use dependent	•Depends on treatment and use	•Site specific	• Facility should be designed to ensure that no reclaimed water reaches potable water supply aquifers. • See Section 3.6 for more information. • For injection projects, filtration and disinfection may be needed to prevent clogging. • See Section 2.4.3 for recommended treatment reliability.
Indirect Potable Reuse Groundwater recharge by spreading into potable aquifers.	•Site specific •Secondary[4] and disinfection[4] (min.) May also need filtration[5] and/or advanced wastewater treatment[16]	•Site specific •Meet drinking water standards after percolation through vadose zone	Includes, but not limited to, the following: •pH - daily •Coliform - daily •Cl₁ residual-continuous •Drinking water standards - quarterly •Other[17] - depends on constituent	•2000 ft. (600 m) to extraction wells. May vary depending on treatment provided and site-specific conditions.	• The depth to groundwater (i.e., thickness of the vadose zone) should be at least 6 feet (2m) at the maximum groundwater mounding point. • The reclaimed water should be retained underground for at least 1 year prior to withdrawal. • Recommended treatment is site-specific and depends on factors such as type of soil, percolation rate, thickness of vadose zone, native groundwater quality, and dilution. • Monitoring wells are necessary to detect the influence of the recharge operation on the groundwater. • The reclaimed water should not contain measurable levels of pathogens after percolation through the vadose zone.[12]

Table 5 EPA Guidelines for Water Reuse (Continued)

Types of Reuse	Treatment	Reclaimed Water Quality[2]	Reclaimed Water Monitoring	Setback Distances[3]	Comments
Indirect Potable Reuse Groundwater recharge by injection into potable aquifers	•Secondary [4] •Filtration [3] •Disinfection [6] •Advanced wastewater treatment [16]	Includes, but not limited to, the following: •pH = 6.5 - 8.5 •≤ 2 NTU [8] •No detectable fecal coli/100 mL [8,10] •1 mg/L Cl$_2$ residual (min.) •Meet drinking water standards	Includes, but not limited to, the following: •pH - weekly •Turbidity - continuous •Coliform - daily •Cl$_2$ residual - continuous •Drinking water standards - quarterly •Other [17] depends on constituent	•2000 ft. (600 m) to extraction wells. May vary depending on site specific conditions.	• The reclaimed water should be retained underground for at least 1 year prior to withdrawal. • Monitoring wells are necessary to detect the influence of the recharge operation on the groundwater. • Recommended quality limits should be met at the point of injection. • The reclaimed water should not contain measurable levels of pathogens at the point of injection. [12] • A higher chlorine residual and/or a longer contact time may be necessary to assure virus inactivation.
Augmentation of surface supplies	•Secondary [4] •Filtration [3] •Disinfection [6] •Advanced wastewater treatment [16]	Includes, but no limited to the following: •pH = 6.5 - 8.5 •≤ 2 NTU [8] •No detectable fecal coli/100 mL [8,10] •1 mg/L Cl$_2$ residual (min.) [11] •Meet drinking water standards	Includes, but not limited to the following: •pH - daily •Turbidity - continuous •Coliform - daily •Cl$_2$ residual - continuous •Drinking water standards - quarterly •Other [17] depends on constituent	•Site specific	• Recommended level of treatment is site specific and depends on factors such as receiving water quality, time and distance to point of withdrawal, dilution and subsequent treatment prior to distribution for potable uses. • The reclaimed water should not contain measurable levels of pathogens. [12] • Reclaimed water should be clear, odorless, and contain no substances that are toxic upon ingestion. • A higher chlorine residual and/or a longer contact time may be necessary to assure virus inactivation.

Table 5 EPA Guidelines for Water Reuse (Continued)

Footnotes

1. These guidelines are based on water reclamation and reuse practices in the U.S., and they are especially directed at states that have not developed their own regulations or guidelines. While the guidelines should be useful in many areas outside the U.S., local conditions may limit the applicability of the guidelines in some countries. It is explicitly stated that the direct application of these suggested guidelines will not be used by AID as strict criteria for funding.

2. Unless otherwise noted, recommended quality limits apply to the reclaimed water at the point of discharge from the treatment facility.

3. Setback distances are recommended to protect potable water supply source from contamination and to protect humans from unreasonable health risks due to exposure to reclaimed water.

4. Secondary treatment processes include activated sludge processes, trickling filters, rotating biological contactors, and many stabilization pond systems. Secondary treatment should produce effluent in which both the BOD and SS do not exceed 30 mg/L.

5. Filtration means the passing of wastewater through natural undisturbed soils or filter media such as sand and/or anthracite.

6. Disinfection means the destruction, inactivation, or removal of pathogenic microorganisms by chemical, physical, or biological means. Disinfection may be accomplished by chlorination, ozonation, other chemical disinfectants, UV radiation, membrane processes, or other processes.

7. As determined from the 5-day BOD test.

8. The recommended turbidity limit should be met prior to disinfection. The average turbidity should be based on a 24-hour time period. The turbidity should not exceed 5 NTU at any time. If SS is used in lieu of turbidity, the average SS should not exceed 5 mg/L.

9. Unless otherwise noted, recommend coliform limits are median values determined from the bacteriological results of the last 7 days for which analyses have been completed. Either the membrane filter or fermentation tube technique may be used.

10. The number of fecal coliform organisms should not exceed 14/100 mL in any sample.

11. Total chlorine residual after a minimum contact time of 30 minutes.

12. It is advisable to fully characterize the microbiological quality of the reclaimed water prior to implementation of a reuse program.

13. The number of fecal coliform organisms should not exceed 800/100 mL in any sample.

14. Some stabilization pond systems may be able to meet this coliform limit without disinfection.

15. Commercially processed food crops are those that, prior to sale to the public or others, have undergone chemical or physical processing sufficient to destroy pathogens.

16. Advanced wastewater treatment processes include chemical clarification, carbon adsorption, reverse osmosis and other membrane processes, air stripping, ultrafiltration, and ion exchange.

17. Monitoring should include inorganic and organic compounds, or classes of compounds, that are known or suspected to be toxic, carcinogenic, teratogenic, or mutagenic and are not included in the drinking water standards.

there have not been any documented cases of viral disease resulting from the reuse of wastewater in the U.S.

For golf course irrigation, the guidelines recommend that the wastewater receive secondary treatment, filtration, and disinfection to produce reclaimed water containing no detectable fecal coliform organisms per 100 mL of sample, as determined from the results of the last 7 days for which analyses have been completed. Due to the high quality of the reclaimed water, setback distances to prevent public contact with the water are not deemed necessary; however, a setback distance of 50 feet (15 m) is recommended between the irrigated area and potable water supply wells to prevent contamination of the potable supply from chemical constituents that may be present in reclaimed water. The guidelines also recommend that a chlorine residual of 0.5 mg/L or greater be maintained in the distribution system to reduce odors, slime, and bacterial regrowth.

References

1.   American Water Works Association. 1990. Recommended practice for backflow prevention and cross-connection control. Manual M14, American Water Works Association, Denver, CO.

2.   California Dept. of Health Services. 1986. Guidelines for use of reclaimed water. Calif. Dept. of Health Services, Office of Drinking Water, Berkeley, CA.

3.   California Dept. of Health Services. 1988. Policy statement for wastewater filtration plants with direct filtration. Calif. Dept. of Health Services, Environmental Management Branch, Sacramento, CA.

4.   California State Water Resources Control Board. 1990. California municipal wastewater reclamation in 1987. Calif. State Water Resources Control Board, Sacramento, CA.

5.   Crook, J. 1989. Viruses in reclaimed water.

p. 231-237.*In* Proc. 63rd Annual Technical Conference, Florida Section American Water Works Assoc., Florida Pollution Control Assoc., and Florida Water & Pollution Control Operators Assoc., November 1989, St. Petersburg Beach, FL.

6. Crook, J. 1990. Water reclamation. *In* Encyclopedia of Physical Science and Technology 1990 Yearbook, Academic Press, Inc., San Diego, CA.

7. Crook, J. 1991. Regulatory issues associated with reuse practices throughout the world. p. 225-239. *In* Proc. 1991 Annual AWWA Conference. June 1991, Philadelphia, PA.

8. Engineering-Science. 1987. Monterey wastewater reclamation study for agriculture: Final report. Prepared for the Monterey Regional Water Pollution Agency by Engineering-Science, Berkeley, CA.

9. Feachem, R.G., H. Bradley, H. Garelick, and D.D. Mara. 1983. Sanitation and disease—health aspects of excreta and wastewater management. Published for The World Bank by John Wiley & Sons, Chichester, England.

10. Florida Dept. of Environmental Regulation. 1983. Land application of domestic wastewater effluent in Florida. Florida Dept. of Environmental Regulation, Tallahassee, FL.

11. Florida Dept. of Environmental Regulation. 1990. 1990 reuse inventory. Florida Dept. of Environmental Regulation, Tallahassee, FL.

12. Florida Dept. of Environmental Regulation. 1991. Guidelines for preparation of reuse feasibility studies for applicants having responsibility for wastewater management. Florida Dept. of Environmental Regulation, Tallahassee, FL.

13. Hireskorn, R.A., and R.A. Ellison. 1987. Sea pines public service district implements a comprehensive reclaimed water system. p. 309-331. *In* Proc. of Water Reuse Symposium IV, Au-

gust 1987, Denver, CO. American Water Works Association, Denver, CO.

14. Lund, E. 1980. Health problems associated with the re-use of sewage: I. bacteria, II. viruses, II. protozoa and helminths. Working papers. WHO Seminar on Health Aspects of Treated Sewage Re-Use, June 1980, Algers, Algeria.

15. Nat'l. Academy of Sciences-Nat'l. Academy of Engineering. 1973. Water quality criteria 1972. EPA/R3/73/033. Prepared by the Committee on Water Quality Criteria, National Academy of Sciences-National Academy of Engineering for the U.S. EPA, Washington, DC.

16. North Carolina Dept. of Environment, Health and Natural Resources. 1992. Report of proceedings for the proposed revisions to the nondischarge rules—Volume I, summary and conclusions. North Carolina Dept. of Environment, Health and Natural Resources, Environ. Management Commission, Raleigh, NC.

17. Pawlowski, S. 1992. Rules for the reuse of reclaimed water. Paper. Salt River Project Water Reuse Symposium, November 1992, Tempe, AZ.

18. Sanitation Districts of Los Angeles County. 1977. Pomona virus study: Final report. Calif. State Water Resources Control Board, Sacramento, CA.

19. Shuval, H.I., A., Adin, B. Fattal, E. Rawitz, and P. Yekutiel. 1986. Wastewater irrigation in developing countries—Health effects and technical solutions. World Bank Technical Paper Number 51, The World Bank, Washington, DC.

20. State of Arizona. 1987. Regulations for the reuse of wastewater. Arizona Administrative Code, Chapter 9, Article 7, Arizona Dept. of Environmental Quality, Phoenix, AZ.

21. State of California. 1978. Wastewater reclamation criteria. California Administrative Code,

Title 22, Division 4, Calif. Dept. of Health Services, Sanitary Engineering Section, Berkeley, CA.

22. State of Florida. 1988. Water policy. Florida Administrative Code, Chapter 17-40, Florida Dept. of Environmental Regulation, Tallahassee, FL.

23. State of Florida. 1989. Reuse of reclaimed water and land application. Florida Administrative Code, Chapter 17-610, Florida Dept. of Environmental Regulation, Tallahassee, FL.

24. State of North Carolina. 1990. Waste not discharged to surface waters. North Carolina Administrative Code, Section T15A: 02H .0200, North Carolina Dept. of Environment, Health and Natural Resources, Raleigh, NC.

25. State of Texas. 1990. Use of Reclaimed Water. Texas Administrative Code, Chapter 310, Subchapter A, Texas Water Commission, Austin, TX.

26. U.S. Environmental Protection Agency. 1974. Supplement to federal guidelines: Design, operation and maintenance of wastewater treatment facilities. Design Criteria for Mechanical, Electric, and Fluid System and Component Reliability. EPA-430/99-74-001, U.S. EPA, Office of Water Program Operations, Municipal Construction Division, Washington, DC.

27. U.S. Environmental Protection Agency. 1981. Process design manual for land treatment of municipal wastewater. EPA 625/1-81-013. U.S. EPA, Center for Environmental Research Information, Cincinnati, OH.

28. U.S. Environmental Protection Agency. 1992. Guidelines for water reuse. EPA/625/R-92/004, Prepared for the U.S. EPA and U.S. Agency for Int'l. Devlop. by Camp Dresser & McKee Inc. U.S. EPA, Center for Environmental Research Information, Cincinnati, OH.

29. Water Pollution Control Federation. 1983. Wa-

ter reuse—Manual of practice SM-3. Water Pollution Control Federation, Washington, DC.

30. Water Pollution Control Federation. 1989. Water reuse (2nd Ed.)—Manual of Practice SM-3. Water Pollution Control Federation, Alexandria, VA.

31. Westcot, D.W. and R.S. Ayers. 1984. Irrigation water quality criteria. *In* G.S. Pettygrove and T. Asano (ed.) Irrigation With Reclaimed Municipal Wastewater—A Guidance Manual. Calif. State Water Resources Control Board, Sacramento, CA. Republished1985. Lewis Publishers, Inc., Chelsea, MI.

32. York, D.W. and J. Crook. 1990. Florida's reuse program: Paving the way. Water Environmental and Technology. 2(12):72-76.

Water Rights: Legal Aspects and Legal Liability

Anne T. Thomas *
Partner
Best, Best & Krieger

Water Rights to Reclaimed Water

During the recent drought, reclaimed water has played an increasingly important role in most western states, including California. Under some circumstances, use of reclaimed water can be mandated. Nevertheless, only a few states have clearly defined water rights to reclaimed water. In most states, this is still an unsettled area of the law. Traditional water rights systems do not easily accommodate reclaimed water use, and conflicting policy issues complicate the problem. California is an example of a state which has not yet decided whether to treat reclaimed water as a "new" water source, or to try to accommodate it in the fresh–water system of water rights.

Only a few states have clearly defined water rights to reclaimed water.

The initial issue is ownership of the right as among those entities which supplied the water which passes through the treatment plant. For example, the water which passes through treatment plants in Orange County may be derived from local wells, or from the Metropolitan Water District of California (MWD), or both. The MWD may have acquired the water from the State Department of Water Resources. Any of the supplying entities theoretically could have claimed a contractual or other right to the wastewater, and resold it to other users after discharge from the treatment plant. However, none did, and in 1980 the State Legislature resolved the issue by providing that the owner of a wastewater treatment plant has the ex-clusive right to the treated wastewater as against anyone who had supplied the water discharged into the waste-

*The author acknowledges the assistance of Nguyen (Win) D. Phan in the preparation of this chapter.

91

Among potential supplier claimants, ownership of the wastewater has been allocated to the treating entity.

water collection and treatment system, unless otherwise provided by agreement[1]. Thus, as among potential supplier claimants, ownership of the wastewater has been allocated to the treating entity.

This does not resolve ownership as among that entity and downstream users of the discharged wastewater. In traditional water rights law, one who diverts water from a natural stream within the watershed where the water is used may improve his system and recapture "waste" or "seepage" water from his land, but he may not recapture and reuse "return flow" if it would injure downstream junior water users, who have vested rights in the continuation of stream conditions as they existed at the time of their respective appropriations. If the source of the water is outside the watershed, however, case law in California has held that the importer of such "foreign" water may recapture waste, seepage, and return flow before it leaves his land at any time, even though such waters were previously released into the stream and put to beneficial use by downstream users. The downstream user could acquire a right, good as against other potential downstream users, but it would be a secondary right to that of the original diverter. The secondary user could not compel the original diverter to continue to divert the water, nor to continue to abandon the water to the stream[2].

In traditional water rights law, one who diverts water from a natural stream within the watershed where the water is used may improve his system and recapture "waste" or "seepage" water from his land . . . if it (does not) injure downstream junior water users.

None of the California cases deal with treated wastewater, however, and some attorneys believe that a series of legislative enactments specifically addressing wastewater evince an intent on the part of the Legislature to bring wastewater reuse under the control of the State Water Resources Control Board (SWRCB), and to grant new legal rights to downstream users.

Water Code section 1202(d)[3] defines water available for appropriation to include "water which having been appropriated or used flows back into a stream, lake, or other body of water." Downstream users seeking permits to appropriate treated wastewater argue that this section includes treated wastewater as a return flow, and that their right to appropriate it is not secondary to that of the treatment plant owner.

Water Code sections 1210-1212[4] specifically deal with treated wastewater, but do not clearly differentiate it from "return flow".

Water Code section 1210 states that as against anyone who has supplied the water discharged into the wastewater systems, the owner of the treatment plant shall hold the exclusive right to the treated wastewater. However, "nothing in this article shall affect the treatment plant owner's obligation to any legal user of the discharged waste water." The critical issue is whether the junior water user's right to use the wastewater flow is secondary to that of the treatment plant owner, or takes precedence.

Is wastewater "return flow" back into streams, lakes, or other bodies of water?

Water Code section 1211 requires the owner of the wastewater treatment plant to obtain approval of the State Water Resources Control Board before changing the point of discharge, place of use, or purpose of use. The State Board must use water code provisions (sections 1700 *et seq.*) relating to changes in freshwater diversions in reviewing such changes. The implication of this section is that wastewater is to be treated as subject to ordinary appropriation, and that downstream users may have acquired a legal right to its use which would bar reuse by the treatment plant operator. On the other hand, an approval under Water Code section 1211 is not the same as a priority–dated permit to appropriate water. Therefore, an attorney with the State Water Resources Control Board (SWRCB), the state agency charged with issuing appropriative permits for use of water, has recommended that those wishing to use reclaimed wastewater, including treatment plant operators, obtain an appropriative permit for the water.

In California, those wishing to use reclaimed wastewater, including treatment plant operators, should obtain an appropriate permit for the water.

This would establish a priority date, quantify the right, and cut off any other potential new downstream appropriators. Although the SWRCB has approved some 15 petitions for change in point of discharge, and place and purpose of use for reclaimed wastewater, it has not yet issued an appropriative permit to a treatment plant operator for reuse of the discharged effluent. The City of Thousand Oaks has filed an application for such a permit, which, if granted, would be the first one issued.

If a treatment plant cannot demonstrate clear ownership of the effluent water, then the availability of the supply to golf course operators is in doubt.

In California, appropriative permits from the SWRCB are required only for use of water in surface streams or lakes. Groundwater use is not subject to the approval of the SWRCB, so treatment plant owners, which have historically discharged wastewater to percolation ponds rather than to a surface stream, should not have conflicts with junior water users, nor be required to seek SWRCB approval in order to reuse discharged wastewater.

The significance of this controversy over water rights for golf course operators is that unless the treatment plant operator can demonstrate clear ownership of the effluent, the availability of the supply may be in doubt. Some states, such as Arizona, have resolved the issue in favor of the treatment plant owners, both to encourage use of reclaimed water, and in recognition of the considerable expense involved in reclaiming the water for beneficial use. Others have not established clear rights to use. The potential purchaser of the water should find out the legal status of reclaimed water in his state, and require assurances of ownership to be included in the purchase contract.

The potential purchaser of the water should find out the legal status of reclaimed water in his state, and require assurances of ownership to be included in the purchase contract.

Mandating Use of Reclaimed Water

Many golf courses are hesitant to use reclaimed water on their courses, not only because of potential liability issues, but also because of public perceptions.

Many golf courses are hesitant to use reclaimed water on their courses, not only because of potential liability issues, but also because of public perceptions. Given the choice, many would prefer to continue using fresh water, even at a higher cost. That choice may not be available in California as a result of legislative declarations and the effect of Article 10, section 2 of the California Constitution[5], which states that no one has a right to water which is wasted, or is not put to reasonable use in a reasonable way. Because of the scarcity of fresh water, the California Legislature has enacted a statute declaring that the use of potable domestic water for non–potable uses, including golf courses, is a waste or an unreasonable use of the water if reclaimed water is available and certain conditions are met. The conditions are that the reclaimed water is of adequate quality for the use, considering all relevant factors; that it is furnished at a reasonable cost to the user, which is comparable to, or less than the cost of potable domestic water; that the use will not be detrimental to public health; and that the use will not adversely affect downstream water rights, will not degrade water quality, and is determined not to be injurious to plant life, fish, and wildlife. Each condition must be found by the SWRCB, at a noticed hearing, on a user by user basis[6]. If all of the conditions are found to exist, the golf course operator can be compelled to use reclaimed water, or be refused water service.

California has enacted a statute declaring that golf courses use reclaimed water if it is available.

Other laws further the state policy of encouraging use of reclaimed water. Suppliers of water may require as a condition of new service that water reclamation devices be installed to save water[7]. The State Water Resources Control Board may require new applicants for appropriative rights to adopt a water reclamation program if the requirements are physically and financially feasible and are appropriate to the particular situation[8]. In a water shortage emergency, as declared by the legislative body of the public agency water supplier, service to new users can be refused altogether, if necessary to preserve sufficient supplies to meet domestic, sanitation, and fire protection needs of existing consumers, and non–essential water uses, including golf courses, can be shut off entirely[9] during the existence of an emergency. A "water shortage emergency" can last for years. Thus, while some bargaining power remains, a water supplier can require golf courses, especially as a condition of new service, to use reclaimed water under appropriate circumstances[10].

Conditions that must be met before reclaimed water is used:
- *adequate quality*
- *reasonable cost*
- *no detrimental health effects*
- *no adverse affects on downstream water rights*
- *maintain downstream water quality*
- *not injurious to plant life, fish and wildlife*

Given the very real possibility of losing water service entirely during droughts, and the fact that the use of reclaimed water on golf courses has yet to produce any reported case of liability, the golf course owner or operator might more profitably allocate its legal resources to drafting a favorable contract than to opposing a reclaimed water requirement, in the absence of compelling economic or other factors.

The Statutory Framework for Use of Reclaimed Water

The goal of the Federal Clean Water Act is that navigable waters were to be "fishable and swim-mable" by 1983 and discharges were to be eliminated by 1990.

Both federal and state laws apply to the use of reclaimed water. In some cases the statutes complement each other, and in other instances they conflict. A typical conflict occurs between an agency, such as the United States Environmental Protection Agency (USEPA), which is charged with reducing pollution in streams, and a state regulatory agency such as the State Water Resources Control Board, which, in addition to stream protection, is also charged with promoting use of reclaimed water. The use of reclaimed water may require flexibility in applying the Clean Water Act requirements for certain streams, to allow use of reclaimed water. However, the USEPA prefers a single, nationwide standard to be applied, without local variances. This can make use of reclaimed water impossible in some areas.

The Federal Clean Water Act[11] is based on federal authority over "navigable" waters. It regulates discharge of contaminants into navigable waters, as defined (including wetlands), through National Pollution Discharge Elimination System (NPDES) permits. The USEPA has jurisdiction, but the Act is administered and enforced by California Regional Water Quality Control Boards. The goal of the Act is that navigable waters were to be "fishable and swimmable" by 1983 and discharges were to be eliminated by 1990. It is obvious that we are many years away from these goals.

The Federal Safe Drinking Water Act regulates drinking water and controls health risks by establishing and enforcing maximum contaminant levels for various compounds in drinking water.

The Porter–Cologne Water Quality Control Act of 1969[12] is a California state law administered by the Regional Water Quality Control Boards in nine regions, and by the State Water Resources Control Board. It regulates "point source" discharges into all waters of the state, including groundwater, with the goal of identifying and protecting beneficial uses of water.

The Federal Safe Drinking Water Act[13] regulates drinking water and controls health risks by establishing and enforcing maximum contaminant levels (MCL) for various compounds in drinking water. The 1986 amendments establish well head protection to protect sole source aquifers, and regulate underground injections of hazardous wastes. The USEPA has jurisdiction, but qualified states, including California, monitor and enforce regulations.

California has its own Safe Drinking Water Act[14]. It is administered by the California Department of Health Services (DOHS). Its purpose is to establish safe levels of possible contaminants in the drinking water supply, and to enforce those limits. The DOHS has the power to control the purity, wholesomeness and potability of the public water supply. Generally, a water purveyor is prohibited from distributing water which exceeds state or federal standards for MCLs. No "community" standards are acceptable.

Reclaimed water is defined as water that as a result of treatment of waste, is suitable for a direct beneficial use or a controlled use that would otherwise not occur and is therefore considered a valuable resource.

The California Water Reclamation Law, enacted in 1969 and subsequently amended, together with the Recycling Act of 1991[15], provide the basic statutory framework for reclaimed water use. These acts express state policy to support use of reclaimed water and establish a regulatory scheme for such use. Jurisdiction and enforcement is shared by the California Department of Health Services and Regional Water Quality Control Boards.

Title 22 of the California Administrative Code[16] contains the specific regulations governing use of reclaimed water. Pursuant to Water Code Section 13521, DOHS establishes statewide reclamation criteria for each type of use of reclaimed water, where such use involves the protection of public health. Existing Title 22 regulations were adopted in 1978 and drafts of proposed new regulations have been in circulation for more than six months. These are the basic regulations governing use of reclaimed water in California.

Regulation of Reclamation

Reclaimed water is defined as water that as a result of treatment of waste, is suitable for a direct beneficial use or a controlled use that would otherwise not occur and

Dual plumbing is now required in California when new subdivisions are constructed, so that they will be able to use recycled water when it becomes available.

is therefore considered a valuable resource[17]. Under Water Code Section 13522.5, any person reclaiming or proposing to reclaim water or using or proposing to use reclaimed water within any region must file with the regional board of that region a report containing such information as may be required by the Regional Board. In 1992, a new section of the Water Code was added to waive the requirement of a user report for persons being supplied with reclaimed water if the supplier provides monitoring and reporting and is issued a master reclamation permit, which includes such use[18].

Regional Boards, after consulting with and receiving recommendations from DOHS, prescribe waste water reclamation requirements (WWRRs), if necessary to protect public health, on the reclaimer, or the user, or both. The WWRRs must be in conformance with statewide reclamation criteria established by DOHS. Discharges must ordinarily comply with objectives in applicable water quality control plans, but according to Water Code Section 13523.5[19], a reclamation project cannot be denied solely because a project violates a salinity standard in a water quality control plan.

Dual plumbing is now required in California when new subdivisions are constructed, so that they will be able to use recycled water when it becomes available[20]. This avoids expensive retrofitting, and continues the state policy of encouraging reclaimed water use. Since January 1, 1993, all new lines carrying recycled water are required to be colored purple, or purple wrapped, to help eliminate possible cross connections[21].

Golf course operators should become familiar with federal, state and local regulations to understand their own obligations and to assure themselves that their supplier is in compliance.

Golf course operators should become familiar with the WWRRs governing their use, both to understand their own obligations, and to assure themselves that their supplier is in compliance. Since treatment plants occasionally have upsets which affect the quality and quantity of the effluent, operators would also be prudent to inquire into the experience of the supplier with reclaimed water, and with meeting relevant National Pollutant Discharge Elimination System (NPDES) requirements.

Legal Liability

To date, the discussion of legal liability for injury caused by use of reclaimed water is still hypothetical, inasmuch as there seem to be no reported cases of injury caused by using reclaimed water. No legislative scheme for allocating liabilities has been developed. Thus, principles of general tort and contract law are presumed to apply, though as yet they are untried. Nevertheless, a knowledge of likely legal theories is useful in considering contractual provisions for the purchase of reclaimed water. The following are some theories which could be used to recover damages from injuries caused by reclaimed water.

There seem to be no reported cases of injury caused by using reclaimed water.

Negligence

Negligence is found when there is a duty to exercise some degree of care in the performance of a function; there is a breach of the duty, and injury results as a direct consequence of the breach. Thus, in an early case involving a death from typhoid fever caused by drinking unchlorinated municipal water, a city was held liable for failure to exercise ordinary care in furnishing the impure water[22]. Public agencies will be held to a duty to use reasonable care and diligence in providing pure and wholesome water, or at least water of an adequate quality for the intended use. In the case of reclaimed water, the water, at a minimum, must meet statutory requirements and regulatory conditions. The golf course owner will want to attach the supplier's WWRR's to its contract, and will want the contract to stipulate that the supplier will be liable for any harm caused by its failure to meet those requirements.

Potential legal liabilities:
- *negligence*
- *negligence per se*
- *implied warranty*
- *strict product liability*
- *emotional distress*

Negligence Per Se

Failure to meet statutory requirements for such water may create *negligence per se*, where the very failure to comply with the statute is proof of negligence. This is especially likely to be the case when there are violations of NPDES permits, wastewater discharge requirements, or wastewater reclamation requirements, since these statutes may have been enacted to prevent the harm that occurred, and the plaintiffs are likely to have been part of the class of persons the statute was intended to protect.

Implied Warranty

Likely legal theories are useful in considering contractual provisions for the purchase.

Several cases in jurisdictions other than California concluded that water constitutes "goods" and therefore can be regulated by the theory of implied warranty in the Uniform Commercial Code. UCC section 2–105[23] defines goods as all things identifiable and movable. The courts in Gall v. Allegheny County Health Department (1989) 555 A.2d and Zepp v. Athens (1986) 348 S.E.2d 673[24] held that municipal water is goods and the implied warranty theory of the UCC applied. (But Cf. Coast Laundry, Inc. v. Lincoln City (1972) 9 Or.App. 521; 497 P2d 1224[25] where water was held *not* to constitute goods under the UCC.) If an implied warranty could be imposed on a contract for reclaimed water, the golf course operator would declare the contract breached if the product was not satisfactory for his anticipated use, under the warranty of merchantability or fitness for his particular purpose. The reclaimed water provider would have to know the proposed golf course use, and that the operator was relying on the reclaimed water's fitness for that use, which are both easily established.

Strict Product Liability

Strict liability can be imposed on ultra–hazardous activity and defective product.

Under California law, strict liability can be imposed on (1) ultrahazardous activity and (2) defective product[26]. Because the use of reclaimed water is approved by the State legislature and is not known to be extremely dangerous, it probably is not an ultrahazardous activity subject to strict liability theory. In California, the seller of a *defective* product may be held strictly liable even if the defect is not *"unreasonably dangerous."* Cronin v. J.B.E. Olson Corp. (1972) 8 Cal.3d 121[27].

In a Texas case[28], the plaintiff sued the city for supplying water which was mixed with flammable gas. The court there held specifically that the doctrine of strict liability applied because the water was defective and it created an "unreasonable risk of injury" to the plaintiff. This case may be distinguishable since it involved an unreasonable risk to the plaintiff which may be absent in the use of reclaimed water. However, in California the product only needs to be defective for strict liability to apply. Thus, a court in California might apply strict liability to reclaimed water if it finds the water "defective." Although a product is flawlessly manufactured and designed, it may be "defective" if a reasonably foreseeable use involves a substantial danger not readily recognizable by the ordinary consumer, and the manufacturer fails to provide an adequate warning[29]. Thus, a producer/supplier of reclaimed water is usually required to, and would certainly want to issue appropriate warnings to potential users or buyers of such water to avoid liability under this theory.

Strict liability may apply to reclaimed water if it is found defective.

If strict product liability can be applied to reclaimed water, then a manufacturer, a retailer or a wholesaler can be liable[30]. Since strict liability extends to any "foreseeable user or consumer" of the product, a golfer who contracted diseases due to the use of reclaimed water for the golf course could recover under this theory, against both the supplier and the golf course.

Emotional Distress

Many cases attempting to link an environmental disease with a proximate cause, e.g., lung cancer to cigarettes, have failed because the plaintiff has been unable to prove that the product caused the disease. An easier case may be made, however, of emotional distress caused by fear of cancer while using a product. This has been recognized as a "toxic tort" by courts and could be applied to users of reclaimed water[31]. In this case, however, adequate notice of the use of reclaimed water, both to employees and to golfers, should prevent recovery, since neither group is required to come upon the course, and both groups would be given the opportunity to avoid exposure.

Emotional distress caused by fear . . . has been recognized as a "toxic tort" by courts.

Contractual Provisions

The golf course operator should have its legal counsel review the contract carefully to ensure that the risks are appropriately allocated.

Most contracts for reclaimed water divide responsibility for the quality and quantity of reclaimed water at the point of delivery. The supplier is responsible for meeting statutory requirements and for operation and maintenance of the plant and delivery systems, and the golf course operator is responsible for providing notices required by law, for installation of appropriate equipment, for adequate employee training and supervision, and for operation of the irrigation system. The golf course operator should have its legal counsel review the contract carefully to ensure that the risks are appropriately allocated. Both parties should acquire and maintain adequate insurance for the life of the contract. In addition, the operator should make its water needs, including both quality and seasonal quantity clear to the supplier, and indicate its reliance on the supplier providing reclaimed wastewater satisfactory for that purpose. With adequate preparation, reclaimed wastewater can be a low cost, reliable asset for golf course developers everywhere.

References

1. Cat. Wat. Code section 1210.

2. Stevens v. Oakdale Irrigation District (1939) 13 Cal.2d 343; Los Angeles v. San Fernando (1975) 14 Cal.3d 199; Crane v. Stevinson (1936) 5 Cal.2d 387.

3. Cal. Wat. Code section 1202(d)

4. Cal. Wat. Code sections 1210-1212.

5. Article 10, section 2 of the California Constitution.

6. Cal. Water Code section 13550.

7. Cal. Water Code section 1009.

8. Cal. Code Regs. 780, tit. 23, Section 23.

9. Cal. Water Code sections 350, 353.

10. Cal. Water Code sections 13550, 13551.

11. Federal Clean Water Act. Federal Water Pollution Prevention and Control Act as amended; Pub.L. 92-500, 33 U.S.C. 1251-1387.

12. Cal. Water Code sections 13000 *et seg.*

13. 42 U.S.C. 300f, 1974 Pub.L. 93-523.

14. Health & Saf. Code sections 4010 *et seg.*

15. California Water Code sections 13500-13554.3; California Water Code sections 13575-13577.

16. Cal. Code Regs., tit. 22.

17. Cal. Water Code sections 13050(n).

18. Cal. Water Code sections 13269, 13523.1.

19. Cal. Water Code section 13523.5.

20. Cal. Water Code sections 13552.2, 13552.3.

21. Health & Saf. Code section 4049.54.

22. Ritterbusch v. City of Pittsburgh (1928) 205 Cal. 84.

23. UCC section 2-105.

24. Gall v. Allegheny County Health Department (1989) 555 A.2d 786 and Zepp v. Athens (1986) 348 S.E.2d 673.

25. Cf. Coast Laundry, Inc. v. Lincoln City (1972) 9 Or.App. 521; 497 P2d 1224.

26. Witkin, Summary of California Law, Vol. 6, Section 1217, et seg.

27. Cronin v. J.B.E. Olson Corp. (1972) 8 Cal.3d 121.

28. Moody v. Galveston (Tex. Civ. App. 1975) 524 S.W.2d 583.

29. Witkin, Summary of California Law, Vol. 6, Section 1265.

30. Vandenmark v. Ford Motor Co. (1964) 61 Cal.2d 256; Witkin, Vol. 6, Section 1279; Witkin, Vol. 6, Section 1282.

31. Potter v. Firestone Tire & Rubber Co. (1990) 232 Cal.App.3d 1114.

Chapter 3 Wastewater Quality, Treatment and Delivery Systems

Wastewater Quality and Treatment Plants

Ali Harivandi, Ph.D.
Turfgrass Specialist
University of California

Design of Delivery Systems for Wastewater Irrigation

Andrew Terrey
Water Resources Specialist
City of Phoenix, Phoenix, AZ

Monitoring Concerns and Procedures for Human Health Effects

Marylynn Yates, Ph.D.
Groundwater Quality Specialist
University of California—Riverside

Wastewater Quality and Treatment Plants

Ali Harivandi, Ph.D.
University of California
Cooperative Extension
Hayward, CA

Introduction

Irrigation water quality is determined by interpretations of a water's chemical analysis.

Irrigation water quality plays a major role in the successful management of turfgrasses and other landscape plants. Of prime importance are the effects of irrigation water on plant–soil–water relations and on the soil's chemical and physical properties, particularly as these factors relate to the quality of turfgrasses and landscape plants. Assuming proper irrigation practices, irrigation water quality is determined by interpretations of a water's chemical analysis.

The possibility of irrigating turfgrasses with reclaimed water (wastewater) is increasingly attractive, especially in highly populated areas. This is due to such factors as water shortages, the rising cost of fresh water and the availability of better quality reclaimed waters.

The chemical and biological constituents of reclaimed water can vary considerably.

The chemical and biological constituents of reclaimed water, however, can vary considerably, and determine the success of using such water for turf and landscape irrigation. Constituents of primary concern appear in Table 1. The composition of untreated and treated reclaimed water—i.e., both actual types and amounts of physical, chemical and biological constituents—depends upon the original composition of the municipal water supply, the number and type of commercial and industrial establishments, and the nature of the residential communities that discharge into the supply[2, 16, 20]. The reclaimed water quality data (Table 2) from selected wastewater treatment plants in California illustrate the considerable variation possible between wastewater treatment plants[2].

In most cases, reclaimed water which has gone through a secondary and/or advanced treatment process is suitable for golf course irrigation. In fact, there may be relatively few instances where treated municipal wastewater quality is poor enough to preclude its use for irrigation. Nevertheless, because reclaimed waters do contain impurities, each situation must be carefully evaluated for the water's possible long–term effects on soils and plants.

All irrigation waters, including reclaimed water, contain varying quantities of soluble salts and other materials. These may include sodium (Na), potassium (K), calcium (Ca), magnesium (Mg), chloride (Cl), bicarbonate (HCO$_3$), sulfate (SO$_4$), nitrate (NO$_3$), and boron (B)[18,19]. Some of these elements may accumulate in the soil to levels injurious to plants; a laboratory chemical analysis can therefore anticipate potential problems posed by use of irrigation water. The most important factors for judging water quality as determined in the analysis are: (1) total salt content; (2) sodium levels (permeability); (3) toxic ion levels; (4) bicarbonate; and (5) pH. In the case of reclaimed water, the nutrient content (e.g., nitrogen, phosphorous, potassium) of the water is often significant and should be evaluated along with the foregoing factors[1,3].

The following discussion focuses on the most common quality problems associated with reclaimed water.

All irrigation waters include varying quantities of
- *sodium (Na)*
- *potassium (K)*
- *calcium (Ca)*
- *magnesium (Mg)*
- *chloride (Cl)*
- *bicarbonate (HCO$_3$)*
- *sulfate (SO$_4$)*
- *nitrate (NO$_3$)*
- *boron (B)*

Important factors for judging water quality:
- *total salt content*
- *sodium level*
- *toxic ion level*
- *bicarbonate*
- *pH*

Total Salt Content

Salinity problems occur when the total quantity of soluble salts in the root zone is too high. Reclaimed waters may be high in salts, and may conceivably lead to soil salt levels intolerable to most turfgrasses and other landscape plants, especially in heavy soils.

Reclaimed waters may be high in salts.

There is a high negative correlation between salt concentration in the soil solution and rate of plant growth —i.e., plant growth generally decreases as salt levels increase[10].

Table 1. Constituents of Concern in Wastewater Treatment and Irrigation with Effluent Water

Constituent	Measured Parameters	Reason for Concern
Suspended solids	Suspended solids, including volatile and fixed solids	Suspended solids can lead to the development of sludge deposits and anaerobic conditions when untreated wastewater is discharged in the aquatic environment. Excessive amounts of suspended solids cause plugging in irrigation systems.
Biodegradable organics	Biochemical oxygen demand, chemical oxygen demand	Composed principally of proteins, carbohydrates and fats. If discharged to the environment, their biological decomposition can lead to the depletion of dissolved oxygen in receiving waters and to the development of septic conditions.
Pathogens	Indicator organisms, total and fecal coliform bacteria	Communicable diseases can be transmitted by the pathogens in wastewater: bacteria, virus, parasites.
Nutrients	Nitrogen, Phosphorus, Potassium	Nitrogen, phosphorus, and potassium are essential nutrients for plant growth, and their presence normally enhances the value of the water for irrigation. When discharged to the aquatic environment, nitrogen and phosphorus can lead to the growth of undesirable aquatic life. When discharged in excessive amounts on land, nitrogen can also lead to the pollution of groundwater.
Stable (refractory) organics	Specific compounds (e.g., phenols, pesticides, chlorinated hydrocarbons)	These organics tend to resist conventional methods of wastewater treatment. Some organic compounds are toxic in the environment, and their presence may limit the suitability of the wastewater for irrigation.

Table 1. Constituents of Concern in Wastewater Treatment and Irrigation with Effluent Water (continued)

Constituent	Measured Parameters	Reason for Concern
Hydrogen ion activity	pH	The pH of wastewater affects metal solubility as well as alkalinity of soils. Normal range in municipal wastewater is pH = 6.5 to 8.5, but industrial waste can alter pH significantly.
Heavy metals	Specific elements (e.g., Cd, Zn, Ni, Hg)	Some heavy metals accumulate in the environment and are toxic to plants and animals. Their presence may limit the suitability of the wastewater for irrigation.
Dissolved solids	Total dissolved solids, electrical conductivity, specific elements (e.g., Na, Ca, Mg, Cl, B)	Excessive salinity may damage some crops. Specific ions such as chloride, sodium, boron are toxic to some crops. Sodium may pose soil permeability problems.
Residual chlorine	Free and combined chlorine	Excessive amounts of free available chlorine (0.05 mg/L) may cause leaf–tip burn and damage some sensitive crops. However, most chlorine in reclaimed wastewater is in a combined form, which does not cause crop damage. Some concerns are expressed as to the toxic effects of chlorinated organics in regard to groundwater contamination.

Source: Asano, T., et al, 1984.

Table 2. Effluent–quality Data from Selected Advanced Wastewater Treatment Plants in California[a,b]

Quality Parameter	Plant Locations					
	Long Beach	Los Coyotes	Pomona	San Ramone Dublin	City of Livermore	Simi Valley CSD
Total nitrogen	-	-	-	-	-	19
NH$_3$-N	3.3	13.6	11.4	0.1	1.0	16.6
NO$_3$-N	15.4	1.1	3	19.0	21.3	0.4
Org-N	2.2	2.5	1.3	0.2	2.6	2.3
Total -P	-	-	-	-	-	-
Ortho-P	30.8	23.9	21.7	28.5	16.5	-
pH (unit)	-	-	-	6.8	7.1	-
Total coliform bacteria, MPN/100 mL[c]	-	-	-	2	4	-
Cations:						
Ca	54	65	58	-	-	-
Mg	17	18	14	-	-	-
Na	186	177	109	168	178	-
K	16	18	12	-	-	-
Anions:						
SO$_4$	212	181	123	-	-	202
Cl	155	184	105	147	178	110
Electrical dS m^{-1} conductivity	1.35	1.44	1.02	1.27	1.25	-
Total dissolved solids	867	87	570	-	-	585
Sodium absorption ration	5.53	4.94	3.37	4.6	5.7	-
Boron (B)	0.95	0.95	0.66	-	1.33	0.6

[a]Advanced wastewater treatment in these plants follows high–rate secondary treatment and includes addition of chemical coagulants (alum + polymer) as necessary, followed by filtration through sand or activated carbon media.
[b]Values expresses in mg L^{-1}, except as noted.
[c]Most probably number/100 milliliter of water sample.

Source: Asano, T., et al, 1984.

A high soil salt level may affect plants by osmotic inhibition of water absorption thus making water less available to plants (physiological drought). Where salinity is very high, grass roots wilt and plants eventually die. Nutritional imbalances and mineral toxicities also may occur at high salinity levels due to competition from salt ions[9,10].

Water salinity is most commonly measured by electrical conductivity (EC_w), and is reported either in millimhos per centimeter (mmhos cm^{-1}) or decisiemens per meter (dS m^{-1}); the two unit's values are equal; only their names differ. Some laboratories, however, report water salinity in micromhos per cm (μmhos cm^{-1}). Therefore, the following equivalences are useful to note:

$$1 \text{ dS m}^{-1} = 1 \text{ mmhos cm}^{-1} = 1000 \text{ μmhos cm}^{-1}$$

Water salinity also may be reported as Total Dissolved Solids (TDS) in parts per million (ppm) or milligrams per liter (mg L^{-1}). The relationship between a water's electrical conductivity (EC_w) and its total dissolved solid (TDS) is approximated as[3,19]:

$$EC_w \text{ (in mmhos cm}^{-1} \text{ or dS m}^{-1}) \times 640 = TDS \text{ (in mg L}^{-1} \text{ or ppm)}$$

As a general rule, salinity problems are associated with irrigation waters with EC_w's greater than 0.75 dS m^{-1}. Severe problems are caused by waters with EC_w's greater than 3.0 dS m^{-1}, and water with EC_w above this level is not recommended for irrigation[3].

The extent of salt intake and its consequent effects on plant growth is related to the salt concentration of the soil solution. The growth of most turfgrasses is not significantly affected by soil salt levels below 3 dS m^{-1}, while at salt levels of 3 to 10 dS m^{-1}, the growth of most turfgrasses is restricted, and above 10 dS m^{-1} only very salt-tolerant turfgrasses can survive[9]. This categorization provides only the most general guidelines to the effect of salinity on turfgrass growth. Pronounced differences among turfgrass species and cultivars in their tolerance of both individual salts and total salinity

Water salinity is most commonly measured by electrical conductivity (EC_w), and is reported either in millimhos per centimeter (mmhos cm^{-1}) or decisiemens per meter (dS m^{-1}).

Water salinity also may be reported as Total Dissolved Solids (TDS) in parts per million (ppm) or milligrams per liter (mg L^{-1}).

As a general rule, salinity problems are associated with irrigation waters with EC_w's greater than 0.75 dS m^{-1}.

necessitates evaluation of each species with a given water and soil salinity combination. However, Table 3 provides a general guide to individual turfgrass salt tolerances.

Although several grasses will tolerate high salt levels, their quality may not be acceptable for all uses. These grasses also may be difficult to propagate or weedy in nature. It is not unusual for a desirable turfgrass species grown on a marginal soil high in soluble salts to be invaded by less desirable grasses. A common example is Kentucky bluegrass turf contaminated by a more salt tolerant bermudagrass. In addition, Table 3 does not indicate possible differences in salt tolerance during stand establishment. Such differences can be significant. Most of the literature on plant salt tolerance determines salt sensitivity by reduction in growth or yield[11,19]. Given the wide range of turfgrass use and management, such reductions may not be of paramount concern. Consequently, the classification presented in Table 3 is based on the estimated potential of turfgrasses to produce fair quality turf under good management.

It is not unusual for a desirable turfgrass species grown on a marginal soil high in soluble salts to be invaded by less desirable grasses.

Because salinity often varies within a site, grasses of various tolerance levels may be grown successfully in combination. For example, mixtures of sod–forming, cool season grasses such as Kentucky bluegrass are often mixed with more salt tolerant perennial ryegrass or alkaligrass.

Stress caused by climatic conditions and soil characteristics other than salinity, make grasses more sensitive to salt stress.

Even relatively salt tolerant grasses can have their tolerance reduced by adverse conditions. Stress caused by climatic conditions and soil characteristics other than salinity, make grasses more sensitive to salt stress[10]. Although a soil chemical test might indicate that a grass is likely to perform adequately in a given soil, other stresses not measured by that test might reduce the grass' performance. Because of the large number of variables present under field conditions, research on salt tolerance done under closely controlled conditions in greenhouses, growth chambers, and laboratories can only provide general guidelines. Also, salt tolerant grasses may provide only a temporary solution for saline sites: if soil drainage and precipitation are insufficient for leaching, salt can eventually build to levels intolerable for all grasses[9].

Table 3. Estimated Salt Tolerance of Common Turfgrasses

Cool–season turfgrass		Warm season–turfgrass	
Name	Rating[a]	Name	Rating[a]
Alkaligrass (*Puccinellia* spp.)	T	Bahiagrass (*Paspalum notatum* Fluegge)	MS
Annual bluegrass (*Poa annua* L.)	S	Bermudagrass (*Cynodon* spp.)	T
Annual ryegrass (*Lolium multiflorum* Lam.)	MS	Blue grama [*Bouteloua gracilis* (H.B.K.) Lag. ex steud.]	MT
Chewings fescue (*Festuca rubra* L. spp. *commutata* Gaud.)	MS	Buffalograss [*Buchloe dactyloides* (Nutt.) Engelm.]	MT
Colonial bentgrass (*Agrostis tenuis* Sibth.)	S	Centipedegrass [*Eremochloa ophiuroides* (Munro) Hackel]	S
Creeping bentgrass (*Agrostis palustris* Huds.)	MS	Seashore paspalum (*Paspalum vaginatum* Swartz.)	T
Creeping bentgrass cv. Seaside	MT	St. Augustine grass [*Stenotaphrum secundatum* (Walter) Kuntze]	T
Creeping red fescue (*Festuca rubra* L. spp. *rubra*)	MS	Zoysiagrass (*Zoysia* spp.)	MT
Fairway wheatgrass [*Agropyron cristatum* (L.) Gaertn.]	MT		
Hard Fescue (*Festuca longifolia* Thuill.)	MS		
Kentucky bluegrass (*Poa pratensis* L.)	S		
Perennial ryegrass (*Lolium perenne* L.)	MT		
Rough bluegrass (*Poa trivialis* L.)	S		
Slender creeping red fescue cv. Dawson (*Festuca ruba* L. spp. *trichophylla*)	MT		
Tall Fescue (*Festuca arundinacea* Schreb.)	MT		
Western wheatgrass (*Agropyron smithii* Rydb.)	MT		

[a]The rating reflects the general difficulty in establishment and maintenance at various salinity levels. It in no way indicates that a grass will not tolerate higher levels with good growing conditions and optimum care. The ratings are based on soil salt levels (EC_e) of: Sensitive (S) = <3 dS m^{-1}, moderately sensitive (MS) = 3-6 dS m^{-1}, moderately tolerant (MT) = 6-10 dS m^{-1}, tolerant (T) = 10 dS m^{-1}.
Source: Harivandi, et al, 1992

In general, water picks up varying amounts of inorganic salts in a cycle of municipal/industrial use. Depending on the initial salt content of the water, this addition could make the resulting water unsuitable for turfgrass irrigation. For example, if the original potable water containing 600 ppm salt picks up 300 ppm additional salts, the resulting reclaimed water with 900 ppm would pose a potential hazard to turf, especially on heavy clay soils.

Sodium Hazard (Permeability)

Sodium's indirect effect on plant growth through its deteriorating effect on soil structure is of greatest concern to turf and landscape managers.

Sodium concentration is another important factor in reclaimed water quality. Although high levels of sodium may accumulate and become toxic to grasses and other plants, it is sodium's indirect effect on plant growth through its deteriorating effect on soil structure which is of greatest concern to turf and landscape managers[3,19,21].

High reclaimed water sodium content causes deflocculation of the soil clay particles, which in turn severely reduces both soil aeration and water infiltration and percolation[14,19,21]. In other words, permeability is reduced when waters containing high levels of sodium are used for irrigation. The effect of an irrigation water source on the relative permeability of a soil is often expressed as Sodium Adsorption Ratio (SAR), the ratio of sodium ion concentration to that of calcium plus magnesium. The following formula calculates the SAR of a water where values for sodium (Na), calcium (Ca), and magnesium (Mg) are given in miliequivalents per liter (meq L^{-1}):

The effect of a irrigation water on the relative permeability of a soil is often expressed as SAR (Sodium Adsorption Ratio).

$$SAR= \frac{Na}{\sqrt{\frac{Ca+Mg}{2}}}$$

If the values for Ca, Mg and Na are given in ppm (mg L^{-1}) in a soil analysis report then the following formula converts meq L^{-1} values:

meq L^{-1} x Equivalent Weight = ppm (mg L^{-1})

Equivalent weights for Na, Ca and Mg are 23, 20 and 12.2, respectively. Generally, a high water SAR (SAR>9) can cause severe permeability problems when applied to fine textured (clay) soils over a period of time. In coarse textured (sandy) soils, permeability problems are less severe and an SAR of this magnitude can be tolerated. Golf greens constructed with high sand content root zone mixes with good drainage, for example, can be maintained using high SAR irrigation waters.

A high water SAR (SAR>9) can cause severe permeability problems when applied to fine textured (clay) soils over a period of time.

In soil, sodium related impermeability problems are measured as the Exchangeable Sodium Percentage (ESP). Sodic soils contain excess sodium ions relative to calcium and magnesium ions. Sodium does not usually cause direct injury to turfgrasses, which among landscape plants are relatively sodium tolerant[9]. Generally, however, if the ESP exceeds 15, a turf stand may be damaged by soil impermeability to water and air[19]. Typical symptoms of reduced permeability include waterlogging, slow water infiltration, crusting and/or compaction, poor aeration, weed invasion, and disease infestation. All of these effects are detrimental to plant growth and development.

In soil, sodium related impermeability problems are measured as the Exchangeable Sodium Percentage (ESP).

Reduced soil permeability also can occur when the salt content of irrigation water is very low (below 0.5 dS m^{-1})[3,14,19]. Water with minimal salt content reduces permeability by dissolving calcium and other soluble particles which disperse and fill soil pore space.

If the ESP exceeds 15, a turf stand may be damaged by soil impermeability to water and air.

Salts and sodium do not act independently in the plant environment. It has been shown that the effects of sodium on soil particle dispersion (and therefore impermeability) are counteracted by high electrolyte (soluble salts) concentration; therefore, the soil sodicity hazard cannot be assessed independently of salinity[3,17]. The combined effect of salinity (electrical conductivity) and sodicity (SAR) on the degree of impermeability caused by a given water is shown in Table 4. Again, the table provides only general guidelines for interpretation of irrigation water quality. Soil properties, irrigation management, climatic conditions, the plant's salt tolerance, and cultural practices all play major roles in the effects caused by irrigation water containing given levels of salt and sodium.

Soil sodicity hazard cannot be assessed independently of salinity.

Table 4. Guidelines for Interpretations of Water Quality for Irrigation

Potential Irrigation Problem	Units	None	Slight to Moderate	Severe
Salinity				
EC_w TDS	dS m^{-1}	<0.7	0.7 - 3.0	>3.0
	mg L^{-1}	<450	450 - 2000	>2000
Infiltration (SAR and EC_w affects infiltration rate of water into the soil. Evaluate using EC_w and SAR together)				
SAR = 0 - 3 and EC_w =		>0.7	0.7 - 0.2	<0.2
SAR = 3 - 6 and EC_w		>1.2	1.2 - 0.3	<0.3
SAR = 6 - 12 and EC_w		>1.9	1.9 - 0.5	<0.5
SAR = 12 - 20 and EC_w		>2.9	2.9 - 1.3	<1.3
SAR = 20 - 40 and EC_w		>5.0	5.0 - 2.9	<2.9
Specific Ion Toxicity				
Sodium (Na)				
root absorption	SAR	<3	3 - 9	>9
foliar absorption	meq L^{-1}	<3	>3	
	mg L^{-1}	<70	>70	
Chloride (Cl)				
root absorption	meq L^{-1}	<2	2 - 10	>10
	mg L^{-1}	<70	70 - 355	>355
foliar absorption	meq L^{-1}	<3	>3	
	mg L^{-1}	<100	>100	
Boron (B)	meq L^{-1}	<1.0	1.0 - 2.0	>2.0
Miscellaneous Effects				
Bicarbonate (HCO$_3$),	meq L^{-1}	<1.5	1.5 - 8.5	>8.5
unsightly foliar deposits	mg L^{-1}	<90	90 - 500	>500
pH		Normal Range 6.5 - 8.4		
Residual chlorine	mg L^{-1}	<1.0	1 - 5	>5

Source: Adapted from Westcot and Ayers 1984; Farnham, et al, 1985

Toxic Ions

Reclaimed waters usually contain a wide variety of elements in small concentrations. Problems occur if certain elements accumulate in the soil to levels toxic to turfgrasses and other plants. The most common toxicities can occur due to accumulation of boron, chloride, and/or sodium[9,12]. Although chloride and sodium are not particularly toxic to turfgrasses, most trees and shrubs are quite sensitive to both[3,6]. A chloride content of 10 meq L^{-1} (355 ppm), or a sodium content of 3 meq L^{-1} (70 ppm) can cause severe foliar damage to sensitive ornamental plants (Table 4). Boron, which enters effluent water through the use of soaps and detergents, is a more likely cause of toxicity in turfgrasses[12]. The major symptom of this toxicity is necrosis at leaf tips, where boron concentrates. Since turfgrasses are mowed regularly and accumulated boron is thus continuously removed from the leaves, most regularly–mowed turfgrasses can tolerate irrigation water high in this element. However, the high boron content of poor reclaimed water poses a greater toxicity problem for trees, shrubs, ground covers, etc. Most landscape plants show injury when irrigated with water containing more than 2.0 mg L^{-1} of boron (Table 4).

Reclaimed water may contain heavy metals, primarily copper, nickel, zinc and cadmium. Zinc and copper are usually beneficial to turf and other landscape plants, while nickel and cadmium are of concern only if, at a later date, the land will be used for other agricultural purposes (e.g., crop production). The National Academy of Sciences has recommended that any water used for irrigation should contain no more than 0.01 mg L^{-1} of cadmium, 0.2 mg L^{-1} of copper, 0.2 mg L^{-1} of nickel, and 2.0 mg L^{-1} of zinc (Table 5). Most secondary reclaimed waters will meet these standards, but continual monitoring is essential.

The most common toxicities can occur due to accumulation of boron, chloride, and/ or sodium.

The major symptom of boron toxicity is necrosis at leaf tips.

Reclaimed water may contain heavy metals, primarily copper, nickel, zinc and cadmium.

Table 5. Recommended Maximum Concentrations of Trace Elements in Irrigation Water

Element	Recommended Maximum Concentration (mg L^{-1})	Remarks
Al aluminum	5.0	Can cause non-productivity in acid soils (pH <5.5), but more alkaline soils at pH >7.0 will precipitate the ion and eliminate any toxicity.
As arsenic	0.10	Toxicity to plants varies widely, ranging from 12 mg L^{-1} for Sudangrass to less than 0.05 mg L^{-1} for rice.
Be beryllium	0.10	Toxicity to plants varies widely, ranging from 5 mg L^{-1} for kale to 0.5 mg L^{-1} for bush beans.
Cd cadmium	0.01	Toxic to beans, beets and turnips at concentrations as low as 0.1 mg L^{-1} in nutrient solutions. Conservation limits recommended due to its potential for accumulation in plants and soils to concentrations that may be harmful to humans.
Co cobalt	0.05	Toxic to tomato plants at 0.1 mg L^{-1} in nutrient solutions. Tends to be inactivated by neutral and alkaline soils.
Cr chromium	0.10	Not generally recognized as an essential growth element. Conservative limits recommended due to lack of knowledge on its toxicity to plants.
Cu copper	0.20	Toxic to a number of plants at 0.1 to 1.0 mg L^{-1} in nutrient solutions.
F fluoride	1.0	Inactivated by neutral and alkaline soils.
Fe iron	5.0	Not toxic to plants in aerated soils, but can contribute to soil acidification and loss in availability of essential phosphorus and molybdenum. Overhead sprinkling may result in unsightly deposits on plants, equipment and buildings.
Li lithium	2.5	Tolerated by most crops up to 5 mg L^{-1}; mobile in soil. Toxic to citrus at low concentrations (<0.075 mg L^{-1}). Acts similarly to boron.
Mn manganese	0.20	Toxic to a number of crops at a few-tenths to a few mg L^{-1}, but usually only in acid soils.

Table 5. Recommended Maximum Concentrations of Trace Elements in Irrigation Water (continued)

Element	Recommended Maximum Concentration (mg L^{-1})	Remarks
Mo molybdenum	0.01	Not toxic to plants at normal concentrations in soil and water. Can be toxic to livestock if forage is grown in soils with high concentrations of available molybdenum.
Ni nickel	0.20	Toxic to a number of plants at 0.5 mg L^{-1} to 1.0 mg L^{-1}; reduced toxicity at neutral or alkaline pH.
Pl lead	5.0	Can inhibit plant cell growth at very high concentrations.
Se selenium	0.02	Toxic to plants at concentrations as low as 0.025 mg L^{-1} and toxic to livestock if forage is grown in soils with relatively high levels of added selenium. An essential element to animals but in very low concentrations.
Sn tin	—	Effectively excluded by plants; specific tolerance unknown.
Ti titanium	—	(See remarks for tin)
W tungsten	—	(See remarks for tin)
V vanadium	0.10	Toxic to many plants at relatively low concentrations.
Zn zinc	2.0	Toxic to many plants in widely varying concentrations; reduce toxicity at pH >6.0 and in fine textured or organic soils.

Source: Wescot and Ayers 1984

Bicarbonate

Reclaimed water's bicarbonate (HCO_3) content can also affect soil permeability and must be evaluated along with the sodium, calcium and magnesium content of both soil and water. The bicarbonate ion may combine with calcium and/or magnesium and precipitate as calcium and/or magnesium carbonate. As calcium and magnesium precipitate out of the soil solution, the SAR of that solution, and consequently the ESP of the soil increases[3]. When dealing with reclaimed water, many analytical laboratories adjust the calculated SAR to include a more correct estimate of the calcium that can be expected to remain in the soil water after an irrigation. This adjusted SAR—expressed as Adj. SAR—reflects the water content of calcium, magnesium, sodium, and bicarbonate, as well as its total salinity.

A water's bicarbonate hazard also can be evaluated in terms of Residual Sodium Carbonate (RSC).

Table 4 indicates tolerable levels of bicarbonate in irrigation waters. A water's bicarbonate hazard also can be evaluated in terms of Residual Sodium Carbonate (RSC)[5]. The following formula calculates the RSC:

$$RSC = (HCO_3 + CO_3) - (Ca + Mg)$$

in which concentrations of carbonate ion (CO_3^{2-}), bicarbonate ion (HCO_3^-), calcium ion (Ca^{2+}) and magnesium ion (Mg^{2+}) are expressed in meq L^{-1}. Generally, waters with RSC values of 1.25 meq L^{-1} or lower are safe for irrigation; those with RSC values of 1.25 to 2.5 are marginal; and those with RSC values of 2.5 meq L^{-1} are probably not suitable for irrigation[5,19].

Generally, waters with RSC values of 1.25 meq L^{-1} or lower are safe for irrigation.

In addition to affecting soil permeability, water high in bicarbonate can increase soil pH to undesirable levels.

pH (Hydrogen Ion Activity)

The pH of reclaimed water is seldom a direct problem by itself, but a very high or very low pH warns that the water needs evaluation for other constituents. The use of pH in evaluating water quality is analogous to the use of body temperature when diagnosing an illness in a human: just as abnormal temperatures indicate illness but do not specify which one, abnormal pHs indicate a problem of some kind exists. The desirable soil pH for most plants is 5.5 to 7.0. The desirable irrigation water pH, however, ranges from 6.5 to 8.4[18,19]. Water with a pH outside the desirable range should be carefully analyzed and evaluated.

The desirable soil pH for most plants is 5.5 to 7.0. The desirable irrigation water pH, however, ranges from 6.5 to 8.4.

Irrigating with high bicarbonate reclaimed water will gradually increase soil pH and lead to moderately alkaline conditions (pH = 7.0 to 8.5). Trace element deficiencies often occur in plants grown in soils with high pH. In the western U.S., for example, naturally high soil pH is one of the major factors causing iron deficiency chlorosis.

Irrigating with high bicarbonate reclaimed water will gradually increase soil pH and lead to moderately alkaline conditions (pH = 7 to 8.5).

Nutrient Content

The nutrient value of reclaimed water is an important economic consideration. Reclaimed water can be high in nutrients (Table 2). Nitrogen, phosphorus, and potassium, all of which are quite beneficial in turfgrass management programs, are primary nutrients present in reclaimed water. The economic values of these nutrients could be substantial and of great value to golf course managers.

Nitrogen, phosphorus, and potassium, all of which are quite beneficial in turfgrass management programs, are primary nutrients present in reclaimed water.

Even if the quantities of nutrients in a given reclaimed water are low, because they are applied on a regular basis, the nutrients are efficiently used by the turfgrass. In most cases turf and other landscape plants will obtain all the phosphorous and potassium they need, and a large part of their nitrogen requirement, from reclaimed water. Sufficient micronutrients are also supplied by most reclaimed waters[4,15].

Managing Water Quality Problems

Salinity Problems

Where salinity is a potential problem due to poor quality reclaimed water, the following management practices should be considered:

Managing salinity problems
- *blending poor and good quality water*
- *planting salt tolerant grasses*
- *applying extra water to leach excess salts*
- *modifying soil profile*
- *installing artificial drainage*

- Blending poor quality water with a less salty water. Frequently a poor quality water can be used for irrigation if better quality water is also available. The two waters can be pumped into a reservoir to mix and then be used for irrigation[3]. Although the resulting salinity will vary according to the type of salts present and climatic conditions, water quality should improve in proportion to the mixing ratio (e.g., when equal volumes of two waters, one with an EC_w of 5 dS m^{-1} and the other with an EC_w of 1 dS m^{-1} are mixed, the salinity of the blend should be approximately 3 dS m^{-1}). The exact salinity content of the blended water is determined by chemical analysis.

- Planting salt-tolerant grasses (Table 3).

- Applying extra water to leach excess salts. To calculate the amount of extra water needed to leach the salt below the turfgrass root zone (and thus provide a suitable level for a specific turfgrass), the Leaching Requirement (LR) is calculated using the following formula[3]:

$$LR = \frac{EC_w}{5(EC_e) - EC_w}$$

LR is the fraction of the plant's normal water requirement which must be added to the requirement solely for leaching purposes. EC_w is the electrical conductivity of the irrigation water being applied (presumably a saline water) and EC_e is the electrical conductivity of soil extract tolerated by the plant grown (for a specific turfgrass, this is the salinity level indicated in Table 3). For example, if a turfgrass which can tolerate a salinity of 3 dS m^{-1} (EC_e) is irrigated with a water having a salinity of 2 dS m^{-1} (EC$_w$), the leaching requirement would equal:

$$LR= \frac{2}{5(3)-2} =0.15$$

To prevent irrigation water salt from accumulating to hazardous levels for a specific turf species, approximately 15 percent extra water should be applied at each irrigation in addition to the normal watering requirement of that turf. This extra water will continuously leach the salt, assuming drainage is adequate. Obviously, any changes in system input, such as leaching caused by rainfall, can greatly affect the amount of water needed for successful leaching.

- If a hard or clay pan is present, modify the soil profile to improve water percolation and, thus, leaching.

- If shallow water tables are a problem, or the soil does not drain well for any reason, install artificial drainage. Leaching does not occur without drainage.

Sodium Hazard (Permeability)

Treatment of water or soil to correct or prevent permeability problems due to high sodium may include:

- Blending irrigation water with a water lower in sodium content.

- Applying soil amendments such as gypsum (calcium sulfate), sulfur, or sulfuric acid[3,7,14]. These amendments increase the soil supply of calcium either directly, as in the case of gypsum, or indirectly, as in the case of the other two. Sulfur and sulfur–containing materials applied to soils naturally high in calcium may make the calcium more soluble. Once available, the calcium can then replace sodium on clay and organic matter particles, thus preventing excess sodium accumulation. Subsequent leaching will flush out sodium salts accumulated in the root zone. The amount

Managing sodium hazard
- *blending irrigation water*
- *applying gypsum or sulfur*
- *frequent aerification*

123

of sulfur amendment used depends on the SAR of the irrigation water, the quantity of water used, soil texture, and type of amendment. The two major factors in successful sodic soil reclamation are: (a) incorporation of amendments into the soil's top 12 to 24 centimeters, and (b) the presence of internal drainage to facilitate the leaching of sodium ions from the root zone.

- Aerifying frequently.

Toxic Ions

Managing toxic ions
- *blending irrigation waters*
- *more frequent irrigation*
- *applying additional water for leaching*

Practices that reduce the effective concentration of toxic elements include:

- Blending poor quality water with better quality water.

- Irrigating more frequently.

- Applying additional water for leaching. Boron is difficult to leach, requiring, for example, two times the amount of water required to leach soluble salts[13].

Bicarbonate

Managing bicarbonate problems:
- *blending irrigation waters*
- *applying gypsum or sulfur*
- *applying acidifying fertilizer*
- *acidification of water*

Practices that reduce the damaging effects of a water's bicarbonate content include those mentioned earlier to remedy problems caused by a high SAR. The impact of bicarbonate on pH may be reduced by applying acidifying fertilizer (e.g., ammonium sulfate) as part of the regular turf fertilization program. In some cases, high levels of bicarbonate in the water require more drastic measures, such as acidification of the water with sulfuric or phosphoric acids[3,14]. Since acid injection into irrigation water is a specialized procedure, requiring unique measurements and equipment, a turf manager must work closely with a consulting laboratory to determine if acidification is required and, if so, how it may best be accomplished.

pH

Abnormal soil pH may be corrected by application of amendments. Liming materials (oxides, hydroxides or carbonates of calcium and magnesium) are used to increase a soil's pH; i.e., to correct an acidity problem. To lower the pH of soils, acidifying amendments such as elemental sulfur or acidifying fertilizers such as ammonium sulfate are used. The kind and amount of amendments used to correct specific pH problems are determined by factors such as: soil pH, soil texture, soil percent base saturation, fineness of the amendment material and turfgrass species. Working closely with a soil testing laboratory in correcting soil and water pH problems is highly recommended.

Abnormal soil pHs may be corrected by application of amendments:
- *correct acidity problems (low pH) with liming materials*
- *correct alkaline problems (high pH) with elemental sulfur or acidifying fertilizers*

Effluent Water Treatment Processes

An understanding of the procedure used to treat wastewater prior to its reuse reveals some of the problems that may develop due to the use of such water[2,8,20].

Reclaimed water may be primary, secondary, or advanced (tertiary) treated.

Reclaimed water may be primary, secondary, or advanced (tertiary) treated municipal or industrial wastewater. Primary treatment is generally a screening or settling process that removes organic and inorganic solids from the wastewater. As sewage enters the treatment plant, it may float through screens to remove rags, sticks, and other floating objects. Screens vary from coarse to fine and are usually placed in a slanted receptacle so that debris can be scraped off and disposed of. Some treatment plants grind these objects so that they remain in the sewage flow and are removed later in a settling tank.

Primary treatment is generally a screening or settling process that removes organic and inorganic solids.

After the sewage has been screened or ground, it passes into a grit chamber were dense materials such as sand, cinders, and small stones settle to the bottom. Settled material is normally washed and used as landfill.

At this point, sewage still contains undissolved suspended matter which can be removed in either a second tank or a primary clarifier. In either case, this material gradually settles out of the liquid and forms a mass of "raw sludge." Raw sludge is drawn off into a digester, which concentrates it for use as landfill. Liquid remaining in the settling tank is called primary effluent and, if only primary treatment is intended, it may be treated before discharge with chlorine to destroy disease-causing bacteria and reduce odor.

Secondary treatment is a biological process in which complex organic matter is broken down to less complex organic material.

Secondary treatment is a biological process in which complex organic matter is broken down to less complex organic material, then metabolized by simple organisms which are later removed from the wastewater. Secondary treatment can remove up to 90 percent of the organic matter in incoming sewage. The secondary liquid effluent may be chlorinated before release. Currently, reclaimed water used for turf and landscape irrigation must be at least secondary effluent water.

Advanced wastewater treatment consists of processes similar to potable water treatment:
- *chemical coagulation and flocculation*
- *sedimentation*
- *filtration*
- *absorption of compounds by a bed of activated charcoal*
- *reverse osmosis*

Advanced wastewater treatment consists of processes similar to potable water treatment, such as chemical coagulation and flocculation, sedimentation, filtration, absorption of compounds by a bed of activated charcoal, or reverse osmosis. Because advanced treatment usually follows secondary treatment, it is sometimes referred to as "tertiary treatment". These processes provide highly purified waters, especially if followed by chlorination for disinfection. A generalized wastewater treatment flowsheet appears in Figure 1.

References

1. Asano, T. 1981. Evaluation of agricultural irrigation projects using reclaimed water. Agreement 8-179-215-2. Office of Water Recycling, Calif. State Water Resources Control Board, Sacramento, CA.

2. Asano, T., R. G. Smith, and G. Tchobanoglous. 1984. Municipal wastewater: Treatment and reclaimed water characteristics. p. 2:1-2:26. *In* Pettygrove, G. S. and T. Asano (ed.) Irrigation with Reclaimed Municipal Wastewater—A Guidance Manual. Report No. 84-1 wr. Calif. State Water Resources Control Board, Sacramento, CA.

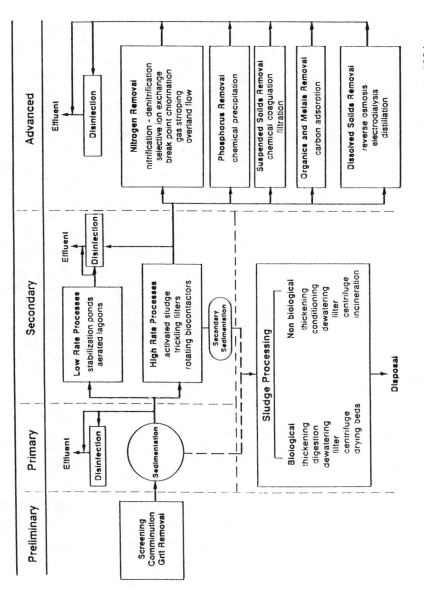

Figure 1. Generalized Flow Sheet for Wastewater Treatment   Source: Asano, T., et al., 1984

3.   Ayers, R. S. and D. W. Westcot. 1985. Water quality for agriculture. FAO Irrigation and Drainage Paper 29 Rev. 1. Food and Agriculture Organization of United Nations.

4.   Broadbent, F. E. and H. M. Reisenauer. 1984. Fate of wastewater constituents in soil and groundwater: Nitrogen and phosphorus. p. 12:1-12:16. *In* Pettygrove, G. S. and T. Asano (ed.) Irrigation with Reclaimed Municipal Wastewater—A Guidance Manual. Report No. 84-1 wr. Calif. State Water Resources Control Board, Sacramento, CA.

5.   Eaton, F. M. 1950. Significance of carbonates in irrigation waters. Soil Sci. 69:123-133.

6.   Farnham, D. S., R. F. Hasek, and J. L. Paul. 1985. Water quality: Its effects on ornamental plants. Univ. of Calif. Coop. Ext. Leaflet 2995. Div. of Agric. Nat. Resources, Oakland, CA.

7.   Harivandi, M. A. 1990. Sulfur, soil pH and turfgrass management. Calif. Turfgrass Culture. 40(1-4): 9-11.

8.   Harivandi, M. A. 1991. Effluent water for turfgrass irrigation. Univ. of Calif. Coop. Ext. Leaflet 21500. Div. of Agric. Nat. Resources, Oakland, CA.

9.   Harivandi, M. A., J. D. Butler, and L. Wu. 1992. Salinity and turfgrass culture. p. 207-229. *In* Waddington, D. V., R. N. Carrow, and R. C. Shearman (ed.) Turfgrass Agron. Monograph No. 32. ASA-C55A-SSSA. Madison, WI.

10.   Levitt, J. 1972. Responses of plants to environmental stresses. Academic Press, New York, NY.

11.   Maas, E. V. 1986. Salt tolerance of plants. Appl. Agric. Res. 1:12-26.

12.   Oertli, J. J., O. R. Lunt, and V. B. Youngner. 1961. Boron toxicity in several turfgrass species. Agron. J. 53:262-265.

13.   Oster, J. D., G. J. Hoffman, and F. E. Robinson.

1984. Management alternatives: Crop, water and soil. Calif. Agric. 38:29-32.

14. Oster, J. D., M. J. Singer, A. Fulton, W. Richardson, and T. Prichard. 1992. Water penetration problems in California soils. Kearney Foundation of Soil Sci., Div. Agric. Nat. Resources, Univ. of California.

15. Page, A. L., and A. C. Chang. 1984. Fate of wastewater constituents in soil and groundwater: trace elements. p. 13:1-13:16. *In* Pettygrove, G. S. and T. Asano (ed.) Irrigation with Reclaimed Municipal Wastewater—A Guidance manual. Report No. 84-1 wr. Calif. State Water Resources Control Board, Sacramento, CA.

16. Pound, C. E. and R. W. Crites. 1973. Characteristics of municipal effluent. p. 49-61. *In* Proc. Recycling Municipal Sludges and Effluents on Land. Nat'l. Assoc. State Univ. and Land-Grant Colleges, Washington, DC.

17. Rhoades, J. D. 1975. Water quality assessment model p. 151-153. *In* Salinity Laboratory Annual Report for 1975. U.S. Salinity Lab. USDA-ARS, Riverside, CA.

18. Rhoades, J. D., and L. Bernstein. 1971. Chemical, physical, and biological characteristics of irrigation and soil water. p. 141-222. *In* L.L. Ciaccio (ed.) Water and Water Pollution Handbook. Vol. 1. Marcel Dekker, New York, NY.

19. Richards, L. A. (ed.) 1954. Diagnosis and improvement of saline and alkali soils. USDA Handb. 60. U.S. Gov. Print. Office, Washington, DC.

20. Tchobanoglous, C. and F. L. Burton. 1991. Wastewater engineering: Treatment, disposal, reuse. McGraw-Hill, New York, NY.

21. Westcot, D. W. and R. S. Ayers. 1984. Irrigation water quality criteria. p. 3:1-3:37. *In* Pettygrove, G. S. and T. Asano (ed.) Irrigation with Reclaimed Municipal Wastewater—A Guidance Manual. Report No. 84-1 wr. Calif. State Water Resources Control Board, Sacramento, CA.

Design of Delivery Systems for Wastewater Irrigation

Andrew Terrey
Irrigation Specialist
City of Phoenix
Water Services Department
Phoenix, AZ

Introduction

Irrigating with reclaimed water is becoming an increasingly popular method for decreasing potable water demand while providing a way to dispose of wastewater without polluting our waterways. In 1991, there were 23 golf courses using effluent in the Phoenix area, with more new and existing facilities going on line each year. Many communities now require new turf facilities, particularly golf courses, to use at least some reclaimed water.

Wastewater irrigation, especially in landscape situations, requires special attention to legal, health, cultural, and environmental concerns.

Wastewater irrigation, especially in landscape situations, requires special attention to legal, health, cultural, and environmental concerns. A report compiled by the Arizona Municipal Water Users Association, however, indicated that Phoenix area golf courses generally made no specific site preparations to accommodate the use of effluent[1]. Interviews with superintendents, on the other hand, revealed that many of them had experienced some significant management problems associated with the use of reclaimed water.

A wastewater delivery system is not just the pipeline that provides a source of water to the turf facility; it is the process by which the effluent is further treated.

A wastewater delivery system is not just the pipeline that provides a source of water to the turf facility; it is the process by which the effluent is further treated. The system must therefore be designed to safely and efficiently dispose of the reclaimed water as well as provide a reliable water source.

This paper discusses some important design considerations regarding delivery systems for reclaimed water irrigation. Some system design criteria cited are required by state and local regulations while other concepts will

130

greatly improve management of the effluent irrigation systems.

Water Reclamation Plants and Delivery Conduits

Every water reclamation plant and delivery system is different. Some schemes consist of a single plant that processes large amounts of wastewater, which is then pumped back to users over an extended area. These systems can be very expensive if the pipelines are extensive. Also, if pipe sizes are not large enough, users may be forced to take water according to a schedule determined by the provider. More popular these days is a smaller "package" plant located on-site or near the turf facility, which processes wastewater for the surrounding community or diverts a portion of it from a trunk sewer.

Reclamation plants
• single plant which processes large amounts of water which is pumped back to users
• small "package" plants located near a turf facility

Delivery method is very important when assessing storage requirements. For example, some plants deliver water on an as-needed basis; others convey water as it is produced—whether the turf facility needs it or not. If the "free-flow" system produces significant quantities of water, adequate storage will be a critical issue, particularly during long periods of rain and cool weather.

During the design phase of the project, the designer should obtain both seasonal and diurnal estimates of the quantity and the quality of water to be delivered. In fact, Arizona Administrative Code (AAC) requires that the producer and the owner of the effluent enter into an agreement which, at a minimum, includes[2]:

Delivery methods
• provided on an as–needed basis
• provided as it is produced

1. The quality and maximum quantity of wastewater to be released for reuse from the plant;

2. the specific reuse(s) for the reclaimed water by the reuser; and

3. the method of disposal by the reuser of any reclaimed wastewater left over from the reuse activity.

131

Firmly establish future estimates of reclaimed water quantity and quality.

These requirements are necessary, but may not be enough. It is important not only to know the present quantity and quality of reclaimed water deliveries, but to firmly establish future estimates since it is the user's responsibility to ensure proper use of the water.

Water reclamation plant personnel are required by law to monitor the quality of the water and keep it within an acceptable range, but these regulations pertain mainly to health considerations. Additional water quality information important to the designer and the superintendent are the following[3]:

Water reclamation plant personnel are required by law to monitor the quality of the water and keep it within an acceptable range.

- Salt Concentration. Salt concentration determines leaching and drainage requirements. It may also indicate potential corrosion problems.

- Sodium Hazard. This ion degrades the soil's ability to aggregate and may lead to drainage problems.

- Bicarbonate Content. Bicarbonate may also lead to sodium problems. Blending effluent with better quality water may be necessary.

- Toxic Ion Concentration. Chlorine and boron are among the toxic ions commonly found in effluent. Although most turf is tolerant of both of these ions, tender trees and shrubs may be harmed.

- pH. The pH of effluent water is generally higher than in its potable water counterpart. The desirable range for irrigation is normally 6.5 to 8.4. Acid injection may be necessary to lower the pH.

Acceptable ranges of these constituents should be established in water delivery contracts, including provisions for dealing with them when designing and building the facility.

Point of Delivery

Some turf facilities use effluent water for only a fraction of their irrigation needs while others are entirely dependent on wastewater. Make sure there is enough water capacity from all sources to meet peak daily needs. Also, if the facility uses a mix of effluent and other water, assess the worst case scenario of what the quality of the mix will be.

Metering is usually required for both billing and regulatory purposes, but is also useful for budgeting and management. Install meters for all water sources at the point of delivery. Depending upon the quality of the effluent, the meter may need to have a screen filter[4].

Metering is usually required for both billing and regulatory purposes, but is also useful for budgeting and management.

Understandably, there should be no cross-connection between potable and reclaimed water lines. The American Water Works Association recommends that effluent distribution lines be buried at least one foot deeper than the domestic water lines and be operated at a lower pressure differential[4]. It may be necessary to install a pressure regulating valve after the meter to accomplish this. If potable water is added to the reclaimed water system, it should be through an approved air-gap separation mechanism[4]. A reduced pressure principle backflow prevention device or a double-check valve should also be installed at the potable water service connection[4]. Check the local building code for the specific requirements.

There should be no cross-connection between potable and reclaimed water lines.

Storage

Water features are an integral part of most golf courses since they provide both irrigation storage and challenging obstacles for golfers. Lakes are critically important when effluent is the sole or primary source of water because they serve as a buffer for daily and seasonal imbalances between supply and demand. Lakes and ponds, however, can be the source of algae, weeds, odors, and health problems if they are not properly designed and managed.

Lakes and ponds can be the source of algae, weeds, odors, and health problems if they are not properly designed and managed.

Lake Size

Lake size varies from facility to facility depending upon seasonal and diurnal water supply and demand. Camp,

Dresser & McKee's *Guidelines for Water Reuse* discusses some important storage considerations[5]. References such as *Irrigation with Reclaimed Municipal Water: A Guidance Manual*[6] and the *Process Design Manual: Land Treatment of Municipal Wastewater*[7] provide methodology for calculating storage requirements.

The size of water features may be affected by local regulations.

The size of water features may also be affected by local regulations. The Arizona Department of Health Services and the Department of Environmental Quality, for example, require facilities that use effluent to have enough storage capacity to meet five days consumptive needs if no means of reuse, discharge, or disposal are available[2]. This mandate is deemed necessary in the event that the ground becomes saturated and cannot hold any more water or if the effluent does not meet minimum water quality standards. Conversely, other regulations may *limit* the size of water features. The Arizona Department of Water Resources Phoenix Active Management Area limits the area of water features not entirely filled with effluent on new golf courses to 0.14 acre per regulation hole[8]. Storage requirements differ from state to state. Check to see what regulations are the most restrictive or if they are in conflict with each other.

Water Quality

Where practical, design lakes to minimize exposure to sunlight:
- *deep and narrow water features*
- *trees, berms and hills which shade*
- *blue dye which absorbs sunlight*

Lake storage should be designed so that it is functional, aesthetically pleasing, and poses minimal health hazards. Maintaining water quality is necessary not only for keeping the golfers satisfied, but also because it is usually required by law. The biggest problem facing lake water quality is the potential for algae and weed growth. Algae cannot always be prevented, but careful attention to lake design can keep it under control.

Two components are necessary for algae growth: nutrients and sunlight. If the effluent is nutrient rich, consider blending it with other water sources, if available. Where practical, design lakes to minimize exposure to sunlight. Reducing sun exposure will also reduce evaporation. Some techniques includes:

1. Deep and narrow water features have less exposed surface area and will not heat up as quickly as shallow, wide lakes.

2. Tall trees, berms and hills around the lake shade the morning and afternoon sun.

3. Blue dye is effective in controlling algae in that is absorbs sunlight[9]. Some courses inject the dye at the effluent inlet.

Although they are not necessarily a design component, aquatic herbivores, like the triploid white amur (sterile grass carp), have been very effective at controlling weeds in central Arizona[9]. Check your state's game and fish department for information about what fish are effective and legal.

Triploid white amur (sterile grass carp), have been very effective at controlling weeds

Adequate circulation and aeration are also necessary for algae and odor control. If practical, install the irrigation system inlet on the opposite end of the lake from the effluent delivery point; the movement of water from the inlet to the outlet will improve circulation. Avoid designing coves or other "dead spots" where the water may stagnate.

There are four ways by which the water can be aerated:

- Fountains. These systems spray water into the air. Though they are fairly effective, they evaporate more water than the other type of aerators.

- Air Injection. Air is blown into the water and allowed to bubble to the surface. This method is highly effective and uses about one-fifth the energy of fountain aerators but may cost up to ten times as much to maintain.

- Water Falls and Streams. Tumbling water over rock or other structures not only aerates the water but can be aesthetically pleasing. Energy costs depend on how much water is pumped and how high it is lifted. Maintenance is about the same as fountain systems.

Adequate circulation and aeration are also necessary for algae and odor control:
- *fountains*
- *air injection*
- *water falls*
- *constructed wetlands*

- Constructed Wetlands. Plants such as cattails and willows are effective both at removing toxins and oxygenating the water. These plants can be located in streams connecting water features or in "out-of-play" regions of lakes. Water needs to circulate around the plants, however, for this type of treatment to be effective.

Irrigation Systems

Irrigation system design considerations:
- *filtration*
- *pH modification*
- *pumps*
- *irrigation control-lers*
- *piping*
- *valves*
- *sprinklers*
- *flush valves*

The irrigation system needs to be designed so that it not only applies water efficiently, but also conforms to legal requirements. Proper design also will improve the quality of the plant material and ensure that the irrigation system will be relatively trouble-free.

Filtration

Probably the greatest problem facing facilities using effluent is clogged equipment due to weeds and algae. Any water pumped from a lake, particularly effluent, should be filtered before entering the irrigation system. Several levels of filtration may be necessary depending upon how susceptible the equipment is to clogging.

All systems that take water in from a lake should have a primary screen (usually 10-30 mesh) at the inlet to prevent large objects from entering the system. Self-cleaning screens that use a jet spray to scour a rotating screen drum prevent the need for divers to manually clean off the filter. Other devices skim water off the top of the lake where the water tends to be cleaner.

Any water pumped from a lake, particularly effluent, should be filtered before entering the irrigation system.

The major filter, usually located at the pumping plant, is used to remove algae and fine debris. Sand filters, wye strainers, and basket or cartage filters can be used depending upon the application and clogging potential. Automatic-discharge strainers that use two separate levels of screening have been very effective filters for reclaimed water[10]. These filtration devices are also effective in controlling Asiatic clams, if used in conjunction with

136

chemical treatment[10]. The flushing mechanism can be connected to either a timer or pressure sensors.

Drip emitters may require additional filtering. A high quality bronze wye strainer with a stainless steel screen of 80 mesh or greater should ensure adequate clog protection[10]. Some emitters may require finer filtration; check with the manufacturer for their recommendation on screen size.

---

pH Modification

Proper pH balance is important not only for plant nutrient uptake, but also to reduce equipment corrosion potential. High or low pH serves as an indication that there may be an ion imbalance in the water[3], which may be improved by mixing the effluent with other water sources or by injecting chemicals.

Many Phoenix-area golf courses use sulfuric acid to adjust the pH of their water. Quite often, however, the equipment is installed after the course is built. Consequently, many injectors are not well integrated into the irrigation systems. Several superintendents have reported severe corrosion problems when applying acid directly at the pump. In one instance, a hydropneumatic tank exploded due to corrosion fatigue, and the shrapnel just missed penetrating the acid holding tank. High acid levels also may damage rubber parts.

To help avoid corrosion problems, acid should be allowed to thoroughly mix with the water before it comes in contact with metal hardware.

To help avoid corrosion problems, acid should be allowed to thoroughly mix with the water before it comes in contact with metal hardware. Some designers recommend injecting acid in the PVC mainline after the pump. It is probably better to adjust the pH before the water enters the lakes or ponds.

The acid injection system should be a "flow-proportion" type that regulates the amount of acid added according to the initial pH and the amount of flow being treated. Do not use models that drip a constant stream of acid whether or not water is being pumped.

137

Pumping Systems

A pumping facility for distributing reclaimed water should be marked to indicate that it is a non-potable source[4]. Also, if there is a potable source of water near the pumping facility, create an air-gap or other acceptable separation to prevent back siphonage into the potable system[4].

If the water has a high salt content, avoid dissimilar metals in the pumping area, like aluminum with iron, that readily corrode when in contact with each other.

If acid and other chemicals are injected near the pump station, stainless steel shafts and bowls are recommended, although they can be very expensive[10]. If the water has a high salt content, avoid dissimilar metals in the pumping area, like aluminum with iron, that readily corrode when in contact with each other. Use high quality brass (red brass) fittings and equipment since they contain smaller amounts of zinc, which corrodes readily when in contact with iron.

Irrigation Controllers

Watering with effluent requires careful irrigation scheduling to avoid ponding and runoff. Most golf courses now have highly sophisticated central irrigation control systems that provide excellent water management capability. When choosing a system, look for features like flow monitoring or flow control, which will shut down the water line in the event of a pipe break. Most controllers allow multiple start times, a necessity for preventing runoff on highly compacted soils.

Look for features like flow monitoring or flow control, which will shut down the water line in the event of a pipe break.

Corrosion is an important consideration when selecting controller cabinets. Use stainless steel cabinets and mount them with stainless steel lag bolts.

Piping

State and local government agencies require that nonpotable waterlines be properly identified and have adequate protection against cross-connection. Local codes should be checked for their requirements. Though laws may differ from location to location, the AWWA *Guidelines for Distribution of Nonpotable Water* should en-

State and local government agencies require that nonpotable waterlines be properly identified.

compass most ordinances. Their recommendations include the following[4]:

1. Nonpotable pipe should be buried at least 1 foot
   deeper than the potable water supply.

2. All buried off-site piping in the nonpotable water
   system, including service lines, should have em-
   bossed lettering, integrally stamped/marked, or be
   installed with warning tape. The color coded
   warning tape (purple is preferred) should be consistent throughout the service area.

*Use "dirty water" or
"contamination proof"
valves.*

3. Hose bibs discharging reclaimed wastewater should
   be secured to prevent any use by the public.

4. Hose bibs discharging reclaimed wastewater should
   be posted with signs reading "Reclaimed Water,
   Do Not Drink", or similar warnings, or be secured
   to prevent access by the public.

5. Quick coupler fittings should be such that inter-
   connection cannot be made between the potable
   and nonpotable system.

Since watering times are usually restricted to when there
is minimal potential for public contact, make sure that
the irrigation system is capable of meeting peak water
demand within a reasonable time window, usually from
7:00 p.m. to 6:00 a.m. (11 hours).

Valves

If the water is properly filtered, clogged valves should
not be a significant problem. To be on the safe side,
many designers now specify "dirty water" or "contamination proof" valves. Some local entities require them
to be color coded.

*To help prevent runoff,
choose models that
have a precipitation
rate that matches, or
is less than, the soil
infiltration rate.*

Sprinklers

Sprinkler selection and location require special consideration. To help prevent runoff, choose models that have
a precipitation rate that matches, or is less than, the soil
infiltration rate. Although valve-in-head sprinklers pro-

vide excellent water control, the water porting device used
to control the valve can become clogged and require more
maintenance than regular heads. Also, sprinklers with
small filters such as pop-up spray heads tend to clog more
easily. Impact heads are generally regarded as being less
clog prone than gear-driven heads. If the water is properly
treated and filtered, however, most any type of sprinkler
can be used.

Impact heads are generally regarded as being less clog prone than gear-driven heads.

State and local authorities are very specific about the
location of sprinkler heads. Arizona law specifically
requires that no spray shall reach privately owned premises, food establishments, or drinking fountains[2]. If the
facility is within a developed area, a buffer zone around
the perimeter that is irrigated with drip irrigation or
bubblers is a good idea.

Flush Valves

Flush valves should be installed in every irrigation system, particularly in low spots and at dead-ends. They
can be used not only to clean out debris, but to purge
air from the pipe.

Drainage and Runoff

Proper drainage is essential for leaching excess salts.

Proper drainage is essential for leaching excess salts.
Good drainage also may be necessary to help prevent
ponding. The type and extent of the drainage system
will depend on the type of soil and the terrain. Consider
installing drains not only on the greens and tees, but also
on fairways and roughs. The best way to encourage good
drainage, however, is to prevent compaction. Pay special
attention to soil preparation after grading and install cart
paths wherever possible.

Proper disposal of drainage water is an important and
difficult design consideration. Some golf course drains
feed back into the lake, which does not really solve the
salt problem since the water is being pumped back onto
the turf. Other systems drain water into large gravel

sumps. In some cases, the water can be discharged back into the sanitary sewer, but this type of system will probably require special permits. Pretreatment of leachate may also be necessary to remove nitrates and other toxic chemicals before it can be discharged into the sewer.

Arizona law requires that all effluent be contained within the premises of the facility and specifically prohibits discharge of water into waterways of the United States without a National Pollutant Discharge Elimination System (NPDES) permit[2]. Design the facility so that it is graded into its interior and has retention basins incorporated into the plan.

Summary

Reclaimed water irrigation system designs are guided by legal requirements, as well as management needs. Before proceeding with the development of a golf course that is going to use reclaimed water, become thoroughly familiar with the federal, state, and local legal mandates. Take a proactive approach and work with government agencies to ensure that the facility is in compliance with all regulations. Also, become familiar with the practical limitations of the water reclamation plant, storage facilities, and the irrigation system, and make sure that they are compatible.

Take a proactive approach and work with government agencies to ensure that the facility is in compliance with all regulations.

Future regulations and restrictions on water reuse may depend on how well the delivery systems are designed and managed today. Good design not only will make the course safe and less maintenance intensive, but will help protect future use of this resource.

References

1. Niemiera, A.X. 1989. Summary of effluent laws and use. Arizona Municipal Water Users Association.

2. Arizona Administrative Code R18-9-702.

3. Kopec, D.M. 1992. Using effluent water for maintaining winter turfgrass. November 1992. Arizona Municipal Water Users Association, Phoenix, AZ.

4. American Water Works Association. California-Nevada Section. 1984. Guidelines for distribution of nonpotable water.

5. Crook, J., D.K. Ammerman, D.A. Okun, and R.L. Matthewes. 1992. Guidelines for water reuse. Camp Dresser & McKee, Inc., Cambridge, MA.

6. Pettygrove, G.S. and T. Asano (ed.) 1985. Irrigation with reclaimed municipal wastewater—a guidance manual. Lewis Publishers, Inc. Chelsea, MI.

7. U.S. Environmental Protection Agency. 1981. Process design manual: Land treatment of municipal wastewater. USEPA 625/1-81-013. EPA Center of Environmental Research Information, Cincinnati, OH.

8. Arizona Dept. of Water Resources. Second Management Plan (1990-2000). The Arizona Groundwater Management Act of 1980. Phoenix Active Management Area, Industrial Conservation Program.

9. Howard, H.F. 1992. Irrigation with effluent. Grounds Maintenance, March, p. 52-58.

10. Coates, G. 1988. Design consideration for irrigation with reclaimed water. Arizona Water & Pollution Control Association Annual Conference, May 1988. Phoenix, AZ.

Monitoring Concerns and Procedures for Human Health Effects

Marylynn V. Yates, Ph.D.
Department of Soil & Environmental Sciences
University of California
Riverside, CA

Introduction

Ground water supplies drinking water to more than 100 million Americans; in rural areas there is an even greater reliance on ground water which comprises up to 95% of the water used[4]. It traditionally has been assumed that ground water is safe for consumption without treatment because the soil acts as a filter to remove pollutants. As a result, private wells generally do not receive treatment,[11] nor do a large number of public water-supply systems. The U.S. Environmental Protection Agency (USEPA) has estimated that approximately 72% of the public water-supply systems in the United States that use ground water do not disinfect[28]. However, the use of contaminated, untreated, or inadequately treated ground water has been the cause of approximately 50% of the waterborne disease outbreaks in this country since 1920[7,8,10,15]. The majority of the outbreaks were caused by pathogenic (disease-causing) microorganisms.

Contaminated, untreated, or inadequately treated ground water has been the cause of approximately 50% of the waterborne disease outbreaks in this country since 1920.

Characteristics of Microorganisms

Bacteria are microscopic organisms, ranging from approximately 0.2 to 10 μm in length. They are distributed ubiquitously in nature and have a wide variety of nutritional requirements. Many types of harmless bacteria colonize the human intestinal tract, and are routinely shed in the feces. One group of intestinal bacteria, the coliform bacteria, historically has been used as an indication that an environment has been contaminated by human sewage. In addition, pathogenic bacteria, such as *Salmonella* and *Shigella*, are present in the feces of infected indi-

Bacteria are microscopic organisms, ranging from approximately 0.2 to 10 μm in length.

viduals. Thus, a wide variety of bacteria can be present in human fecal material, and thus municipal wastewater.

Viruses are obligate intracellular parasites; that is, they are incapable of replication outside of a host organism. They are very small, ranging in size from approximately 20 to 200 nm. Viruses that replicate in the intestinal tract of man are referred to as human enteric viruses. These viruses are shed in the fecal material of individuals who are infected either purposely (i.e., by vaccination) or inadvertently by consumption of contaminated food or water, swimming in contaminated water, or person-to-person contact with an infected individual. More than one hundred different enteric viruses may be excreted in human fecal material[20]. As many as 10^6 plaque-forming units (pfu) of enteroviruses (a subgroup of the enteric viruses) per gram and 10^{10} rotaviruses per gram may be present in the feces of an infected individual[26]. Thus, viruses are present in domestic sewage and, depending on the type of treatment process(es) used, between 50 and greater than 99.99 percent of the viruses are inactivated during sewage treatment.

Viruses are obligate intracellular parasites which range in size from 20 to 200 nm.

A third group of microorganisms of concern in domestic sewage is the protozoan parasites. In general, protozoan parasite cysts (the resting stage of the organism which is found in sewage) are larger than bacteria, although they can range in size from 2 μm to over 60 μm. Protozoan parasites are present in the feces of infected persons; however, they also can be excreted by healthy carriers. Cysts are similar to viruses in that they do not reproduce in the environment, but are capable of surviving in the soil for months or even years, depending on environmental conditions.

Protozoan parasites range in size from 2μm to over 60 μm.

A list of microorganisms that may be transmitted by contaminated water is shown in Table 1.

Sources of Microorganisms

Microorganisms may be introduced into the subsurface environment in a variety of ways. In general, any practice that involves the application of domestic wastewater to the soil has the potential to cause microbiological con-

Table 1. Pathogens Transmitted by Water

Pathogen	Disease
Bacteria:	
Campylobacter jejuni	gastroenteritis
Enteropathogenic *E. coli*	gastroenteritis
Legionella pneumophila	acute respiratory illness
Salmonella	typhoid, paratyphoid, salmonellosis
Shigella	dysentery
Vibrio cholerae	cholera
Yersinia enterocolitica	gastroenteritis
Protozoa:	
Cryptosporidium	diarrhea
Entamoeba histolytica	amoebic dysentery
Giardia lamblia	diarrhea
Naegleria fowleri	meningoencephalitis
Enteroviruses:	
Adenovirus	respiratory illness, eye infection, gastroenteritis
Astrovirus	gastroenteritis
Calicivirus	gastroenteritis
Coxsackievirus A	meningitis, respiratory illness
Coxsackievirus B	myocarditis, meningitis, respiratory illness
Echovirus	meningitis, diarrhea, fever, respiratory illness
Hepatitis A virus	infectious hepatitis
Norwalk virus	diarrhea, vomiting, fever
Poliovirus	meningitis, paralysis
Rotavirus	diarrhea, vomiting

Viruses have been detected in the ground water at several sites practicing land treatment of wastewater.

tamination of ground water. This is because some of the treatment processes to which the wastewater is subjected do not effect complete removal or inactivation of the disease-causing microorganisms present. Goyal[14] isolated viruses from the ground water beneath cropland being irrigated with sewage effluent. Viruses have been detected in the ground water at several sites practicing land treatment of wastewater[17]. The burial of disposable diapers in sanitary landfills may be a means by which pathogenic microorganisms in untreated human waste may be introduced into the subsurface. Vaughn[34] detected viruses as far as 408 m down–gradient of a landfill site in New York. Land application of treated sewage effluent for the purposes of groundwater recharge has also resulted in the introduction of viruses to the underlying ground water[32,33].

Removal by Treatment Processes

Treatment of wastewater can effect from 50% to >99.99% pathogen removal, depending on the treatment processes used.

Treatment of wastewater can effect from 50 percent to greater than 99.99 percent pathogen removal, depending on the treatment processes used. A summary of pathogen removal rates and typical concentrations reported to be present after several stages of sewage treatment is presented in Table 2. It can be seen that even tertiary treatment (consisting of primary sedimentation, trickling filter/activated sludge, disinfection, coagulation, direct filtration, and chlorination) does not remove all pathogens. It is important to consider the infective dose of the organism in relation to the final concentration when assessing the potential public health risk associated with use of reclaimed water. It is relatively unlikely that the two *Salmonella* organisms would cause disease, considering that the infective dose is more than 1000 organisms. On the other hand, the final concentrations of virus and *Giardia* may be sufficiently high to cause some people to become ill if they ingested the water.

Table 2. Pathogen Removal in Treated Sewage Source: Stewart 1990

	Enteric Viruses	Salmonella	Giardia
Infective Dose (no. particles)	1	>1000	25-100
Amount in feces	$10^6 - 10^{10}$ g^{-1}	10^{10} g^{-1}	9×10^6 stool^{-1}
Concentration in raw sewage (no. l^{-1})	$10^5 - 10^6$	5,000 - 80,000	9,000 - 200,000
Removal during:			
Primary treatment[1]			
% removal	50 - 98.3	95.8 - 99.8	27 - 64
No. remaining l^{-1}	1,700 - 500,000	160 - 3,360	72,000 - 146,000
Secondary treatment[2]			
% removal	53 - 99.92	98.65 - 99.996	45 - 96.7
No. remaining l^{-1}	80 - 470,000	3 - 1075	6480 - 109,500
Tertiary treatment[3]			
% removal	99.983 - 99.9999998	99.99 - 99.99999995	98.5 - 99.99995
No. remaining l^{-1}	0.007 - 170	0.000004 - 7	0.099 - 2,951

[1] Primary sedimentation and disinfection
[2] Primary sedimentation, trickling filter/activated sludge, and disinfection
[3] Primary sedimentation, trickling filter/activated sludge, disinfection, coagulation, filtration, and disinfection

Waterborne Disease in the United States

Between 1920 and 1990, a total of 1,674 waterborne disease outbreaks were reported in the United States, involving over 450,000 people and resulting in 1,083 deaths[10,15]. The number of reported outbreaks and the number of associated cases of illness have risen dramatically since 1971, as compared with the period from 1951-1970 (Figure 1)[10]. The increase in reported numbers of outbreaks may be due to an improved reporting system implemented in 1971[6]; however, it is still believed that only a fraction of the total number of outbreaks is reported[19]. Based on survey data from the Centers for Disease Control, it has been estimated that waterborne infections affect 940,000 people and are responsible for 900 deaths every year in the United States[3].

Causative agents of illness were identified in approximately one–half the disease outbreaks during the period from 1971 to 1990 (Table 3). The most commonly identified causative agents were *Giardia*, chemicals, and *Shigellae*. *Giardia lamblia* caused over 18% of the illness associated with waterborne disease outbreaks. Enteric viruses (viral gastroenteritis and hepatitis A) were identified as the causative agents of disease in 8.7% of the outbreaks during this period.

In the 1980's, use of untreated or inadequately treated ground water was responsible for 44% of the outbreaks that occurred in the United States (Figure 1)[10]. When considering outbreaks that have occurred due to the consumption of contaminated, untreated ground water, from 1971 to 1985, sewage was most often identified as the contamination source. In ground water systems, etiologic (disease-causing) agents were identified in only 38% of the outbreaks, with *Shigella* sp. and hepatitis A virus being the most commonly identified pathogens[9]. In over one-half of the outbreaks, no etiologic agent could be identified, and the illness was simply listed as gastroenteritis of unknown etiology. However, retrospective serological studies of outbreaks of acute nonbacterial gastroenteritis from 1976 through 1980 indi-

Between 1920 and 1990, a total of 1,674 waterborne disease outbreaks were reported in the United States, involving over 450,000 people and resulting in 1,083 deaths

When considering outbreaks that have occurred due to the consumption of contaminated, untreated ground water, from 1971 to 1985, sewage was most often identified as the contamination source.

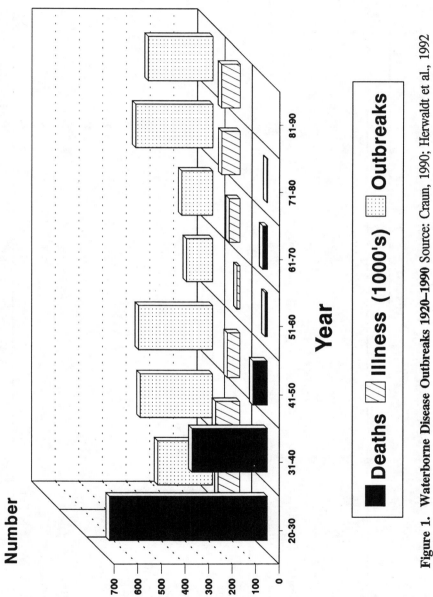

Figure 1. Waterborne Disease Outbreaks 1920–1990 Source: Craun, 1990; Herwaldt et al., 1992

Table 3. Causative Agents of Waterborne Disease Outbreaks, 1971–1990 Source: Craun 1991, Herwaldt et al., 1992

Disease	Outbreaks		Illness	
	Number	% of Total	Number of Cases	% of Total
Gastroenteritis, unknown cause	293	49.66	67,367	47.26
Giardiasis	110	18.64	26,531	18.61
Chemical poisoning	55	9.32	3,877	2.72
Shigellosis	40	6.78	8,806	6.18
Viral gastroenteritis	27	4.58	12,699	8.91
Infectious hepatitis	25	4.24	762	<1
Salmonellosis	12	2.03	2,370	1.66
Campylobacterosis	12	2.03	5,233	3.67
Typhoid fever	5	<1	282	<1
Yersiniosis	2	<1	103	<1
Cryptosporidiosis	2	<1	13,117	9.20
Chronic gastroenteritis	1	<1	72	<1
Toxigenic E. coli	2	<1	1,243	<1
Cholera	1	<1	17	<1
Dermatitis	1	<1	31	<1
Amebiasis	1	<1	4	<1
Cyanobacteria-like bodies	1	<1	21	<1
	564	100	138,247	TOTAL

cated that 42% of these outbreaks (i.e., the 62% for which no etiologic agent was identified) were caused by the Norwalk virus[16]. Thus, it has been suggested that the Norwalk virus is responsible for approximately 23% of all reported waterborne outbreaks in the United States[18].

The difficulty in the isolation of many enteric viruses from clinical and environmental samples probably accounts for the limited number of viruses identified as causes of waterborne disease. As methods for the detection of enteric viruses have improved, so has the percentage of waterborne diseases identified as having a viral etiology[12].

The fact that microorganisms are responsible for numerous waterborne disease outbreaks every year led the U.S. Environmental Protection Agency to re-examine the coliform standard, which has been used to indicate the microbiological quality of drinking water in the United States for more than 75 years. Increasing evidence collected during the past 15 to 20 years suggest that the coliform group may not be an adequate indicator of the presence of pathogenic viruses and parasites in water. For example, in a study of the removal of viruses and indicator bacteria at seven drinking water treatment plants in Canada, none of the bacterial indicators were correlated with the concentration of viruses in the finished water[21].

Increasing evidence collected during the past 15 to 20 years suggest that the coliform group may not be an adequate indicator of the presence of pathogenic viruses and parasites in water.

In 1985, the USEPA proposed Maximum Contaminant Level Goals (MCLGs) for viruses and *Giardia*[27]. These standards are in addition to the standard for the indicator microorganism group, total coliform bacteria. Rather than require public water systems to monitor the water for the presence of these pathogenic microorganisms, the USEPA[28] proposed treatment requirements with the goal that the level of pathogenic viruses and *Giardia* in the treated water would result in a risk of less than one infection in 10,000 persons per year.

Detecting Pathogens in Water

Detecting pathogens, especially viruses and parasites, in surface and ground water requires the collection of very large samples.

Recent advances in recombinant DNA technology have provided new tools that can be used to detect viruses in ground water.

Detecting pathogens, especially viruses and parasites, in surface and ground water requires the collection of very large samples; a typical sample volume is hundreds to thousands of liters. The large sample volume is necessitated by the fact that the concentration of these microorganisms is expected to be very low in non-sewage impacted water. However, due to the fact that one virus or parasite particle may be sufficient to cause infection, even very low concentrations can pose a threat to public health. For example, the World Health Organization[35] has suggested that an acceptable virus concentration in drinking water is zero per 100 to 1000 liters.[23] In the United States, using the USEPA's goal of limiting the risk of increased infections to one in 10,000 persons per year would result in an acceptable virus concentration of 2 viruses per 10 million liters of water.[22].

To sample such large volumes, the water is passed through an electropositive or chemically–treated electronegative filter which adsorbs viruses present in the water. The viruses are then removed from the filter with a small volume of proteinaceous liquid such as beef extract.[1] The concentrated sample may then be analyzed for the presence of viruses using several different techniques. The traditional method involves inoculating the appropriate living cells and waiting for evidence of infection. This process may take up to three weeks to complete. There are several more rapid techniques for detecting viruses in environmental samples, including radioimmunoassay (RIA) and enzyme-linked immunosorbent assay (ELISA). These methods require the use of a specific antibody against the virus of interest. The drawback of such methods is that very high numbers (10,000 to 100,000) of virus particles are required. The number of virus particles that would be expected in a concentrated ground-water sample would be orders of magnitude lower.

Recent advances in recombinant DNA technology have provided new tools that can be used to detect viruses in ground water. Nucleic acid probes can be made to be specific to one virus or a specific group of viruses. The probes have the advantage that results are obtained rapidly; however, they are unable to differentiate between infective and non-infective virus particles. Another recent development that should make virus detection more sensitive is polymerase chain reaction (PCR) technology. PCR enables one to make copies of virus particles present in a sample, thereby amplifying one's ability to detect viruses. These new methods are currently being used for research purposes, and many improvements will have to be made before they could be generally used for detecting viruses and other pathogens in environmental samples. However, the development of this technology will undoubtedly have an major impact on the field of water microbiology.

Risks Associated with Pathogens in Reclaimed Water

There are several possible routes of exposure to pathogens in reclaimed water. One route is through drinking water that has been impacted by reclaimed water. Another route is through contact with plant materials that have been irrigated using reclaimed water. Pathogens can also be contracted by exposure to aerosols generated during spray irrigation with reclaimed water. Finally, persons can become infected indirectly by contact with an individual who has been in direct contact with the reclaimed water. This is termed secondary infection, and can be a significant means of spread of enteric disease. The secondary infection rate may be as high as 90 percent for some enteric viruses (Table 4).

Possible routes of exposure to pathogens:
- *drinking water impacted by reclaimed water*
- *aerosols used during irrigation*
- *contact with infected persons*

Drinking Water

In order for pathogens in reclaimed water to contaminate drinking water, they first must be transported either to surface water or ground water. A discussion of mechanisms controlling the transport and fate of enteric pathogens in the environment is beyond the scope of this paper, but the interested reader is referred to a number of review articles.[4,24,31,36]

Table 4. Secondary Attack Rates of Enteric Viruses

Virus	Secondary Attack Rate (%)
Poliovirus	90
Hepatitis A virus	78
Coxsackievirus	76
Echovirus	40
Norwalk virus	30

Source: Gerba and Rose, 1993

Table 5. Probability of Infection for Enteric Microorganisms

Microorganism	Probability of Infection from Exposure to One Organism
Campylobacter	7×10^{-3}
Salmonella	2.3×10^{-3}
Salmonella typhi	3.8×10^{-5}
Shigella	1.0×10^{-3}
Vibrio cholera classical	7×10^{-6}
Vibrio cholera El Tor	1.5×10^{-5}
Poliovirus 1	1.49×10^{-2}
Poliovirus 3	3.1×10^{-2}
Echovirus 12	1.7×10^{-2}
Rotavirus	3.1×10^{-1}
Entamoeba coli	9.1×10^{-2}
Entamoeba histolytica	2.8×10^{-1}
Giardia lamblia	1.98×10^{-2}

Source: Rose and Gerba, 1991

Based on recent reported waterborne outbreak data, the risk of acquiring an illness from contaminated water in the United States has been estimated to be approximately 4×10^{-5} per year or 2.8×10^{-3} during a lifetime[5]. This risk estimate is probably low due to the fact that many waterborne outbreaks are not reported. There are several reasons for this including the fact that many waterborne pathogens cause gastroenteritis which is generally not severe enough to require medical attention, so unless a large number of people are involved, the outbreak goes unrecognized. Another reason is that reporting of waterborne disease outbreaks is not required in the United States. In addition, it is difficult to assess the number of cases of illness associated with waterborne pathogens because of secondary infection to individuals who were not directly exposed to the contaminated water (Table 4).

The risk of acquiring an illness from contaminated water in the U.S. is approximately 1 in 25,000 per year.

The impacts of very low levels of pathogens in drinking water are difficult to document. However, the fact that only one or two virus or parasite particles may be required to cause infection necessitates an attempt to quantify this risk. The probability of infection resulting from the ingestion of 1 organism has been estimated for several different pathogens using data from dose-response curves derived from human feeding studies and assuming an ingestion rate of 2 liters per day (Table 5). It is important

Table 6. Ratio of Clinical Illness to Subclinical Infections with Enteric Viruses

Virus	Frequency of clinical illness
Polio 1	0.1 - 1
Coxsackie	
A16	50
B2	11 - 50
B3	29 - 96
B4	30 - 70
B5	5 - 40
Echo	
overall	50
9	15 - 60
1	rare - 20
20	33
25	30
30	50
Hepatitis A (adults)	75
Rota	
adults	56 - 60
children	28
Astro (adults)	12.5

Table 7. Mortality Rates for Enteric Microorganisms

Microorganism	Mortality Rate (%)
Polio 1	0.90
Coxsackie	
A2	0.50
A4	0.50
A9	0.26
A16	0.12
Coxsackie B	0.59 - 0.94
Echo	
6	0.29
9	0.27
Hepatitis A	0.6
Rota	
total	0.01
hospitalized	0.12
Norwalk	0.0001
Adeno	0.01

Source: Gerba and Rose, 1993

to realize that infection will not necessarily result in illness in all cases; certain viruses have very low ratios of clinical illness to subclinical infections (Table 6). Infection by certain enteric viruses may result in death; the probability of death is higher for the very young, the very old, and the immunocompromised. Mortality rates for some enteric viruses are shown in Table 7. Because only poliovirus causes a reportable disease (poliomyelitis), the mortality rates represent only hospitalized cases of enterovirus infection.

Using the data from Tables 5 through 7, the probability of infection, illness, and death from a single exposure to a microorganism can be calculated. Regli et al.[22] compared the simple exponential and modified exponential (ß) models with experimental dose–response data. They found that the ß model fit the echovirus 12, poliovirus III, and rotavirus exposure data best, while the exponential model fit the poliovirus 1 data best. The annual risk of infection, disease, and mortality from exposure to three different concentrations of hepatitis A virus in drinking water is shown in Figure 2.

Golf Courses

Asano et al.[2] used the ß model to calculate the risk associated with exposure to viruses on a golf course irrigated using reclaimed wastewater. They assumed that a person golfs twice a week for 30 years, for a total of 3120 days exposure over a lifetime. They also assumed that the golfer is exposed to 1 ml of reclaimed water a day through handling and cleaning golf balls. One day of pathogen inactivation was also included in the calculation, based on the assumption that irrigation occurs during the night, and that golfing occurs on a dry field. Results of their calculations for daily, annual, and lifetime risks for exposure to poliovirus 1 and poliovirus 3 are presented in Figures 3 and 4.

Regulations Pertaining to the Microbiological Quality of Reclaimed Water Reuse

Currently, 27 states have regulations or guidelines for the restricted urban reuse of reclaimed water. These have been compiled by the USEPA,[30] and an abbreviated version is reproduced in Table 8. Restricted urban reuse of reclaimed water is defined as the irrigation of areas in which public access can be controlled such as golf courses, cemeteries, and highway medians.

Currently, 27 states have regulations or guidelines for the restricted urban reuse of reclaimed water.

The regulations vary widely in their stringency. For example, in terms of the microbiological indicators, the reclaimed water may contain from < 1 fecal coliform (in Florida) to 1000 fecal coliforms (in New Mexico). Criteria for monitoring the quality of the reclaimed water also vary widely. South Carolina requires monitoring for viruses, while other states rely on daily or weekly monitoring of indicator bacteria. Several states require that the underlying ground water be monitored to determine the impact, if any, of the application of the reclaimed water. The monitoring generally consists of an upgradient and downgradient well. However, in South Carolina, their regulations specifically state that golf courses must have at least nine monitoring wells per 18 fairways.

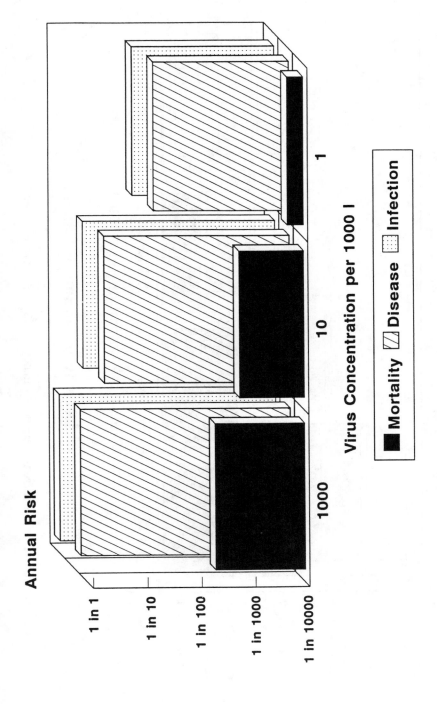

Figure 2. Risk of Infection from Exposure to Poliovirus 1 in Reclaimed Water Source: Gerba and Rose, 1993

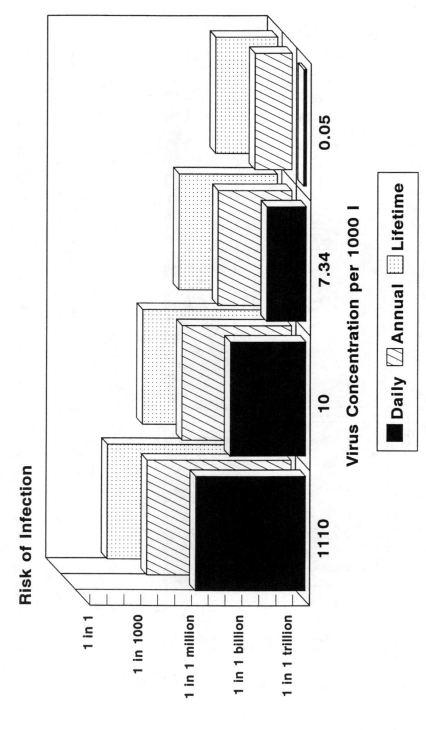

Figure 3. Risk of Infection from Exposure to Echovirus 12 in Reclaimed Water Source: Asano et al., 1992

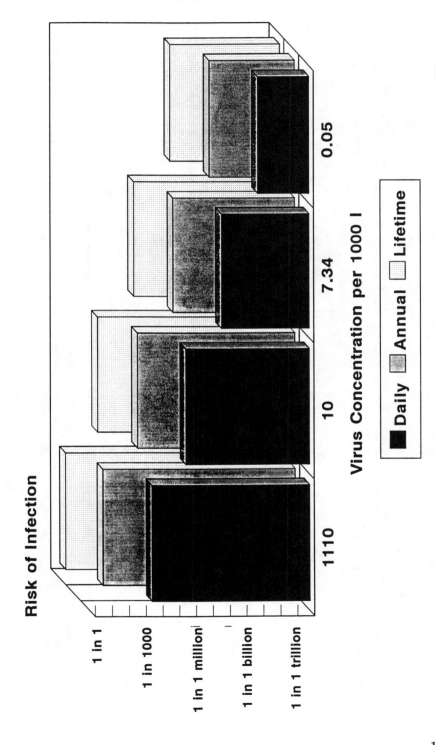

Figure 4. Risk of Infection from Exposure to Poliovirus 3 in Reclaimed Water Source: Asana et al., 1992

Table 8. Current State Regulations and Guidelines for Restricted Urban Reuse of Reclaimed Water

State	Reclaimed Water Quality and Treatment Requirements	Reclaimed Water Monitoring Requirements	Ground-Water Monitoring	Setback Distances
AR	secondary treatment and disinfection		required 1 well downgradient 1 well upgradient 1 well within site	determined case-by-case
AZ	pH - 4.5 - 9 fecal coliform -200/100 ml (median)	pH - 1/month fecal coliform - 1/wk		
CA	disinfected and oxidized total coliform-23/100 ml (median)	total coliform - 1/day		
CO	disinfected and oxidized total coliform-23/100 ml (median)			500' to domestic supply well 100' to any irrigation well
DE	biological treatment 30 mg/l BOD 30 mg/l TSS fecal coliform - 30/100 ml	site specific	required 1 well upgradient 1 well within wetted field area 2 wells downgradient in each drainage basin intersected by site	determined case-by-case
FL	secondary treatment with filtration, high level disinfection and chemical feed facilities;	TSS limit prior to disinfection and monitored daily continuous on-linemonitoring of turbidity	required 1 well upgradient 1 well within reuse site	75' to potable water supply low trajectory nozzles within

FL	20 mg/l CBOD (annual average) 5 mg/l TSS; chlorine residual hourly flow fecal coliform - over 30-day period 75% samples below detection	and chlorine residual; fecal coliform - 1/day for >0.5 mgd annual analysis of primary and secondary drinking water standards	1 well downgradient	100' of outdoor public eating, drinking, and bathing facilities
GA	biological treatment 30 mg/l BOD 30 mg/l TSS fecal coliform - 30/100 ml	site specific	required 1 well upgradient 1 well within wetted field area 2 wells downgradient in each drainage basin intersected by site	determined case-by-case
HI	disinfected and oxidized total coliform-23/100 ml (7 day mean)	total coliform - 1/day	required	spray - 500' to residences or property lines (unless irrigating between 9 pm and 5 am); drip - 5' to residences or property lines (unless irrigating between 9 pm and 5 am); surface - none
ID	disinfected and oxidized total coliform-230/100 ml (median)			
IL	two cell lagoon system with sand filtration and disinfection or mechanical secondary treatment with disinfection		required 1 well upgradient 2 wells downgradient	non-spray application-50' to residential lot, 10' to public road right-of-way, spray - 150' to residential lot

Table 8. Current State Regulations and Guidelines for Restricted Urban Reuse of Reclaimed Water

State	Reclaimed Water Quality and Treatment Requirements	Reclaimed Water Monitoring Requirements	Ground-Water Monitoring	Setback Distances
KS	secondary treatment with disinfection		site specific may be required	none required
MD	30 mg/l BOD 90 mg/l TSS fecal coliform - 3/100 ml pH - 6.5 - 8.5		required 1 well upgradient 2 wells downgradient	spray - 200 ft to property lines, waterways, and roads; 500 ft to housing developments and parks; non-spray - 50 ft to property lines, waterways, roads, housing developments, and parks
MO	secondary treatment disinfected prior to application fecal coliform - 200/100 ml		minimum of 1 well between site and public supply well	150 ft to existing dwellings or public use areas; 50 ft to property lines; 300 ft to potable water supply wells not on property, sinkholes, and losing streams
MT	disinfected and oxidized fecal coliform - 200/100 ml		required when ground-water levels are within 20 ft. of the natural ground surface in the irrigation zone	200 ft to any dwelling

State	Treatment/Water Quality	Monitoring		Setback/Buffer
NE	biological treatment disinfected prior to application	site specific	site specific	
NV	secondary treatment with disinfection no buffer zone - fecal coliform 2.2/ 100 ml (mean); turbidity 3 NTU 100' buffer zone - fecal coliform 23/100 ml; turbidity 5 NTU	samples taken prior to application point and after final treatment		none or 100', depending on level of disinfection
NM	adequately treated and disinfected fecal coliform 1000/100 ml	fecal coliform sample taken at point of diversion to irrigation system		
NC	5 mg/l TSS (monthly average) maximum fecal coliform-1/100 ml	TSS and fecal coliform limits met prior to discharge to detention pond		100 ft vegetative buffer to nearest dwelling
OK	minimum of primary treatment		network of monitoring wells may be necessary and should be evaluated for systems with deep percolation	spray irrigation may require buffer zones to ensure that aerosols are contained on the site
OR	biological treatment and disinfection total coliform-23/100 ml (median)	total coliform - 1/week		surface irrigation - 10' buffer; spray - 70' buffer 100 ft to drinking fountains and areas of food preparation

Table 8. Current State Regulations and Guidelines for Restricted Urban Reuse of Reclaimed Water

State	Reclaimed Water Quality and Treatment Requirements	Reclaimed Water Monitoring Requirements	Ground-Water Monitoring	Setback Distances
SC	secondary treatment with disinfection, chemical addition, and filtration (latter 2 not required for golf courses); BOD $\leq$ 5 mg/l (monthly avg.); SS $\leq$ 5 mg/l (monthly avg.); total coliform - $\leq$ 4/100 ml (monthly avg.); for golf course irrigation - total coliform $\leq$ 200/100 ml	continuous monitoring of turbidity virus monitoring	required 1 well upgradient 2 wells downgradient for golf courses a minimum of 9 wells are suggested for each 18 fairways	75' to residences for golf course irrigation
SD	secondary treatment and disinfection total coliform - 200/100 ml		shallow wells in all directions of ground-water flow from site and not more than 200' outside of site	
TN	treatment requirements on a case-by-case basis; disinfection required; BOD $\leq$ 30 mg/l fecal coliform - $\leq$ 200/100 ml		required	surface - 100' to site boundary, 50' to on-site streams, ponds, and roads; spray - open fields: 300' to site boundary, 150' to on-site streams, ponds, and roads; forested: 150' to site boundary, 75' to on-site streams, ponds, and roads

TX	30 mg/l BOD with treatment using pond system (30-day avg.) 20 mg/l BOD with other treatment (30-day avg.) ; fecal coliform ≤800/100 ml	sampling and analysis 1/month		
UT	advanced treatment BOD ≤ 10 mg/l at any time TSS ≤ 5 mg/l at any time total coliform - ≤ 3/100 ml at any time			
WA	secondary treatment total coliform-20/100 ml (mean)		minimum of one well in each direction of ground water movement from site and one well upgradient of site	
WY	BOD ≤ 10 mg/l (daytime) BOD ≤ 30 mg/l (dusk-dawn) pH - 4.5 - 9.0 ; fecal coliforms - ≤200/100 ml; TDS ≤ 480 mg/l chlorides - ≤213 mg/l spray site			100' buffer zone around spray site

References

1. American Public Health Association. 1985. Standard methods for the examination of water and wastewater. 16th edition. American Public Health Association, Washington, DC.

2. Asano, T., L.Y.C. Leong, M.G. Rigby, and R.H. Sakaji. 1992. Evaluation of the California wastewater reclamation criteria using enteric virus monitoring data. Proc. International Association on Water Pollution Research and Control Sixteenth Biennial Conference and Exhibition, May 1992.

3. Bennett, J.V., S.D. Holmberg, S.F. Rogers, and S.L. Solomon. 1987. Infectious and parasitic diseases. p. 102-114. *In* R.W. Amler and H. B. Dull (ed.) Closing the Gap: The Burden of Unnecessary Illness. Oxford University Press, New York, NY.

4. Bitton, G. and C. P. Gerba. 1984. Groundwater pollution microbiology: The emerging issue. p. 1-7.*In* G. Bitton, and C.P. Gerba (ed.) Groundwater Pollution Microbiology. John Wiley & Sons, New York, NY.

5. Bull, R.J., C. Gerba, and R.R. Trussell. 1990. Evaluation of the health risks associated with disinfection. Crit. Rev. Environ. Control 20:77-113.

6. Craun, G. F. 1985. A summary of waterborne illness transmitted through contaminated groundwater. J. Environ. Hlth. 48:122-127.

7. Craun, G. F. 1986a. Statistics of waterborne outbreaks in the U.S. (1920-1980). p. 73-159. *In* G. F. Craun (ed.) Waterborne Diseases in the United States, Chapter 5. CRC Press, Boca Raton, FL.

8. Craun, G. F. 1986b. Recent statistics of waterborne disease outbreaks (1981-1983). p. 43-69. *In* G. F. Craun (ed.) Waterborne Diseases in the United States, Chapter 4. CRC Press, Boca Raton, FL.

9.   Craun, G. F. 1990. Methods for investigation and prevention of waterborne disease outbreaks, U.S. EPA, Office of Research and Development. Report No. EPA-600/1-90/005a, September, 1990.

10. Craun, G. F. 1991. Causes of waterborne outbreaks in the United States, Water Sci. Technol. 24:17-20.

11. DiNovo, F. and M. Jaffe. 1984. Local Groundwater Protection, Midwest Region. American Planning Association, Chicago, IL.

12. Gerba, C. P. 1984. Strategies for the control of viruses in drinking water. Report to Amer. Assoc. Adv. Sci., Washington, DC.

13. Gerba, C.P. and J.B. Rose. 1993. Estimating viral disease risk from drinking water. p. 117-135. *In* C.R. Cothern (ed.) Comparative Environmental Risk Assessment. Lewis Publishers, Boca Raton, FL.

14. Goyal, S.M., B.H. Keswick, and C.P. Gerba. 1984. Viruses in groundwater beneath sewage irrigated cropland. *Water Res.* 18:299-302.

15. Herwaldt, B. L., G. F. Craun, S. L. Stokes, and D. D. Juranek. 1992. Outbreaks of waterborne disease in the United States: 1989-1990. J. Amer. Water Works Assoc., 84:129 - 135.

16. Kaplan, J. E., G. W. Gary, R. C. Baron, W. Singh, L. B. Schonberger, R. Feldman, and H. Greenberg. 1982. Epidemiology of Norwalk gastroenteritis and the role of Norwalk virus in outbreaks of acute nonbacterial gastroenteritis. Ann. Intern. Med. 96:756-761.

17. Keswick, B. H. and C. P. Gerba. 1980. Viruses in groundwater. Environ. Sci. Technol. 14:1290-1297.

18. Keswick, B.H., T. K. Satterwhite, P. C. Johnson, H. L. DuPont, S. L. Secor, J. A. Bitsura, G. W. Gary, and J. C. Hoff. 1985. Inactivation of Norwalk virus in drinking water by chlorine. Appl. Environ. Microbiol. 50:261-264.

19. Lippy, E. C. and S. C. Waltrip. 1984. Waterborne disease outbreaks—1946-1980: A thirty-five-year perspective. J. Amer. Water Works Assoc., 76:60-67.

20. Melnick, J.L. and C.P. Gerba. 1980. The ecology of enteroviruses in natural waters. *CRC Crit. Rev. Environ. Contr.* 10: 65-93.

21. Payment, P., M. Trudel, and R. Plante. 1985. Elimination of viruses and indicator bacteria at each step of treatment during preparation of drinking water at seven water treatment plants. Appl. Environ. Microbiol. 49:1418-1428.

22. Regli, S. J.B. Rose, C.N. Haas, and C. P. Gerba. 1991. Modeling the risk from *Giardia* and viruses in drinking water. J. Amer. Water Works Assoc. 83:76-84.

23. Rose, J.B. and C.P. Gerba. 1991. Use of risk assessment for development of microbial standards. Wat. Sci. Technol. 24: 29-34.

24. Sobsey, M.D. 1983. Transport and fate of viruses in soils. In Microbial Health Considerations of Soil Disposal of Domestic Wastewaters. pp. 174-191. Cincinnati, OH: U.S. EPA Report No. EPA-600/9-83-017..

25. Stewart, M.H. 1990. Pathogen removal in treated sewage. Metropolitan Water District of Southern California, Water Quality Laboratory. LaVerne, CA.

26. Tyrrell, D. A. and A. Z. Kapikian.1982.Virus Infections of the Gastrointestinal Tract. Marcel Dekker, Inc., New York, NY.

27. U.S. Environmental Protection Agency. 1985. National primary drinking water regulations: synthetic organic chemicals, inorganic chemicals, and microorganisms. Proposed rule. Fed. Reg. 50: 46936-47022.

28. U.S. Environmental Protection Agency. 1989. National primary drinking water regulations: filtration; disinfection; turbidity; *Giardia*

lamblia; viruses; *Legionella*; and heterotrophic bacteria. Final Rule. Fed. Reg., 54:27486.

29. U.S. Environmental Protection Agency. 1990. Strawman regulation for ground water disinfection requirements (GWDR), Office of Drinking Water, Washington, DC. April18, 1990.

30. U.S. Environmental Protection Agency. 1992. Guidelines for Water Reuse. EPA/625/R92/004. Washington, DC.

31. Vaughn.J.M. 1983. Viruses in soils and groundwater. *In* Viral pollution of the environment. G. Berg, ed. CRC Press, Boca Raton, FL. pp. 163-210.

32. Vaughn. J.M. and E.F. Landry. 1977. Data report: An assessment of the occurrence of human viruses in Long Island aquatic system. Dept. of Energy and Environment, Upton, NY: Brookhaven National Laboratory.

33. Vaughn, J.M. and E.F. Landry. 1978. The occurrence of human enteroviruses in a Long Island groundwater aquifer recharged with tertiary wastewater effluents. *In* State of Knowledge in Land Treatment of Wastewater, Vol. 2, Washington, DC: U.S. Govt. Printing Office, pp. 233-245.

34. Vaughn, J.M., E.F. Landry, L.J. Baranosky, C.A. Beckwith, M.C. Dahl, and N.C. Delihas. 1978. Survey of human virus occurrence in wastewater–recharged groundwater on Long Island. Appl. Environ. Microbiol., 36: 47-1.

35. World Health Organization. 1984. Guidelines for drinking water quality. Vol. 1. Recommendations. World Health Organization, Geneva, Switzerland.

36. Yates, M.V. and S.R. Yates. 1988. Modeling microbial fate in subsurface environments. CRC Critical Reviews in Environ. Control. 17:307-344.

Chapter 4   Effects of Wastewater on the Turfgrass/Soil Environment

Irrigation of Turfgrass with Wastewater

Charles Mancino, Ph.D.
Associate Professor, Plant Sciences
University of Arizona

Ian L. Pepper, Ph.D.
Professor, Soil and Water Science
University of Arizona

Coping with Wastewater: Management Strategies and Information Sources

James Moore
Director
USGA Green Section, Mid–Continent Region

Irrigation of Turfgrass with Wastewater

Charles F. Mancino, Ph.D.
Associate Professor
Dept. of Plant Sciences
University of Arizona
Tucson

Ian L. Pepper, Ph.D.
Professor
Dept. of Soil and Water Science
University of Arizona
Tucson

Introduction

Municipal effluent water is ideally suited for turf irrigation, particularly in arid climates. There are several reasons for this including the following: (1) Many arid climates permit continuous growth of turfgrass, allowing year–round utilization of wastewater; (2) The high shoot and root density of turfgrasses has the potential for removing groundwater pollutants from large volumes of wastewater; (3) Large volumes of irrigation water are necessary for the growth of turf on large surface areas, which further increases the volume of effluent which can be cleansed; (4) Plant nutrients, common to wastewater, reduce the need for commercial fertilizer use on turf; (5) Most expanses of irrigated turf are located adjacent to cities, where effluent is produced, thereby minimizing transportation costs; and (6) Potential health problems associated with the use of effluent on turf are less than when effluent is used for irrigation of unprocessed food crops.

Secondary and tertiary treated wastewaters usually contain essential plant macro– and micro–nutrients. These would include nitrogen (N), phosphorous (P), potassium (K), calcium (Ca), magnesium (Mg), sulfur (S), iron (Fe), zinc (Zn), manganese (Mn), copper (Cu), and boron (B). Wastewater is a valuable, but often very costly water source. Cost may range from as little as $98 per ha-m of water to as high as $7,300 per ha-m. In Tucson, AZ, reclaimed wastewater costs approximately $3,746 per ha-m.

Two excellent references that discuss the use of wastewater for irrigation are Pettygrove and Asano[26] and Feigin

Effluent water is suited for turf in arid climates of the South:
- *continuous turf growth*
- *high shoot/root density*
- *large volumes of irrigation required*
- *nutrients in wastewater used by turf*
- *located near cities*
- *turf is a non–food crop*

et al.[11]. However, these references mainly deal with the use of this water for agricultural irrigation. Harivandi[14] focused on turfgrass irrigation with wastewater in a California Cooperative Extension bulletin.

There is a pool of scientific information, mainly from the southwestern U.S., on the use of wastewater for turf irrigation. These studies have examined the influence of wastewater on turf and soil quality and the leaching losses of wastewater constituents from the turfgrass rootzone. The purpose of this paper is to discuss these research findings.

Seedling Establishment

Effluent has been used for the successful production of crops throughout the world[11]. Successful turfgrass irrigation has also been reported[9,16,17,18,21,25] This research has resulted in the increased use of this water source on prestigious golf course facilities in Arizona, California, and Florida.

Effluent irrigation may cause a reduction in the emergence of turfgrass seedlings. This may be due to several characteristics of the irrigation water including high sodium (Na), ammonium–nitrogen (NH_4^+-N), trace element content and, in particular, salinity. Turf seedlings are more sensitive to saline conditions than mature turf[15]. Hayes et al.[16] reported that emergence of 'common' bermudagrass (*Cynodon dactylon* var. dactylon) and perennial ryegrass (*Lolium perenne* L.) seedlings was reduced when irrigated with wastewater. Bermudagrass germination was reduced by 4.8 percent and ryegrass by 5.7 percent when compared to controls. The salinity of the wastewater was deemed to be the cause. However, Hayes et al.[16] observed that seedling establishment was improved due to wastewater irrigation despite fewer initial seedlings being present. Effluent irrigation resulted in greater turfgrass groundcover than potable plots not receiving any fertilizer. Soil analyses prior to planting indicated adequate levels of N, P, and K and no apparent need to apply fertilizer. Wastewater provided a readily

Effluent irrigation may cause a reduction in the emergence of turfgrass seedlings.

175

available source of nutrients to young plants, which have a limited root system. Also, P contained in the wastewater helps to promote establishment[5].

Any reduction in germination due to the wastewater could be compensated for in the field with an increase in seeding rate.

The turfgrass manager should expect seedlings to respond favorably to wastewater that provides some plant nutrients. A standard germination test in petri plates comparing the effects of the wastewater to potable water on germination may be advisable in some cases. A reduction in germination due to the wastewater could be compensated for in the field by an increased seeding rate.

Nutrient Availability

Benefits of Nitrogen

Turfgrass quality under effluent irrigation may be the same as or better than turf irrigated with potable water and N-P-K fertilizer at certain times of the year. Hayes et al.[17] reported better quality during June, July, and August when non–fertilized, effluent–irrigated overseeded plots transitioned from perennial ryegrass back to the bermudagrass. However, quality of both fertilized, effluent–irrigated plots and fertilized potable–irrigated plots was better during other times of the year. During summer months, when the water requirement of the turf was high, enough effluent was being applied to meet the water and N requirements of the turf. In this study, the N requirements of the turf could not be met by the wastewater alone during non–summer months. Therefore, N fertilizer should be applied at these times when wastewater volumes do not provide adequate nitrogen. Hayes et al.[16] reported the best quality wastewater–irrigated turf when 49 kg N ha^{-1} mo^{-1} was applied during fall, winter, and spring months.

Turfgrass quality under effluent irrigation may be the same or better than turf irrigated with potable water and N-P-K fertilizer at certain times of the year.

Heat stress

Additional N from fertilizer, even at low monthly application rates, may result in heat stress on susceptible turfs also irrigated with N–containing wastewaters. Hayes et al.[17] showed that the application of N to effluent irrigated turf at either 16.3, 32.3, or 49 kg N ha^{-1} caused a rapid and severe loss in ryegrass quality during July. However, the underlying bermudagrass responded well to the additional fertilizer N and provided good quality summer turf within one month. Because of the possibility of N–induced heat stress a turfgrass manager should reduce or eliminate N fertilization on N–containing effluent–irrigated turf at least two or three months prior to the stress period. This recommendation is based upon the use of a wastewater which contains about 398 kg N per ha-m water.

Iron Chlorosis

Turf appearance may suffer from Fe chlorosis during certain times of the year due to effluent irrigation. Reference to this type of chlorosis on wastewater–irrigated turf has been made by Anderson et al.[2], Kneebone and Pepper[19], and Hayes et al.[17] All speculated the cause to be high soil pH resulting from wastewater irrigation. The soils all had pH values of 8.0 or greater. Kneebone and Pepper[19] reported hybrid bermudagrass (*C. dactylon x transvaalensis* var. 'Tifway') produced better turf compared to common bermudagrass when both were irrigated with wastewater. Hayes et al.[17] attributed Fe chlorosis on effluent–irrigated, N–fertilized common bermudagrass to two possible causes: (1) N–induced Fe chlorosis in which shoot growth exceeded the ability of the roots to take up Fe from the soil; and (2) the formation of insoluble Fe phosphates occurring because of the P delivered in the wastewater and the high soil pH.[13] Hayes et al.[17] corrected the problem with a foliar application of ferrous sulfate despite the fact that soil and tissue analysis showed adequate levels of Fe.

Turf appearance may suffer from iron chlorosis during certain times of the year due to effluent irrigation.

Salt Effects

Salinity

Any irrigation water, including effluent water, may contain levels of salts and specific ions such as Na, B, and chlorides (Cl) which can be deleterious to plant growth and soil quality.

Harivandi[14] reported the relative salinity tolerances of turfgrasses. Most grasses are tolerant of soil salinities (EC_e) in the range of 3 to 6 dS m^{-1}. The salinity of effluent water often meets the standards set for irrigation waters used for very sensitive crops (EC_w <1 dS m^{-1}). Many reports have discussed the possible detrimental effects of effluent water used to irrigate turf; however, data presented in the scientific literature has not shown this to be the case. To the contrary, most studies have shown effluent to be an excellent source of irrigation water for turfgrass, having a minimal impact on EC_e.[16,17, 21] Therefore, it cannot be assumed that a water is of low quality simply because it is a wastewater. For example, wastewater in Tucson, AZ has an EC_w of 0.6 to 0.9 dS m^{-1}, whereas potable water in Chandler, AZ has an EC_w of 1.1 to 1.9 dS m^{-1}. Thus it is safe to assume that a low quality municipal water will result in a lower quality wastewater. Wastewaters typically contain 100 to 300 mg L^{-1} more salts, predominately NaCl, than the original potable water[6]. This would result in an increase in EC_w of about 0.15 to 0.47 dS m^{-1}. Chandler, AZ wastewater has an EC_w of 1.7 to 2.4 dS m^{-1}.

Work initiated by Hayes et al.[16,17] and continued by Mancino and Pepper[22] found that a Tucson, AZ effluent source (EC_w = 0.65 to 0.91 dS m^{-1}) increased turfgrass soil EC_e from 0.7 dS m^{-1} to about 1.5 dS m^{-1} after 3.5 years. However, similarly irrigated turf plots receiving potable water (0.2 dS m^{-1}) had a final EC_e of 1.3 dS m^{-1}. Thus, the change in EC_e resulting from the use of effluent was actually small.

Although many turfgrasses are quite tolerant of saline soil conditions, this may not be the case with other

Most grasses are tolerant of soil salinities (EC_e) in the range of 3 to 6 dS m^{-1}.

It cannot be assumed that a water is of low quality simply because it is a wastewater.

Although many turfgrasses are quite tolerant of saline soil conditions, this may not be the case with other landscape plants.

landscape plants. The turf manager should confer with local landscape plant authorities to select tolerant annual and perennial herbaceous and woody landscape plants. Actually, this should be done regardless of whether the irrigation source is to be wastewater, especially in arid climates.

Sodium

Sodium is a naturally occurring constituent of waters. Wastewater from municipal sources tends to have a higher Na content because of the use of water softeners. As with salinity, it is difficult to find instances in which the use of wastewater has resulted in the loss of turf due to Na toxicity (or for that matter B or Cl toxicity). The effects of Na, B, and Cl on turf growth were recently reviewed by Harivandi et al.[15]

The inability to grow good turfgrass in soils containing high levels of sodium results not from the direct toxicity of sodium to the plant, but instead from the negative impact that Sodium has on the soil.

Research has shown that soil Na content increases with wastewater irrigation, but this is not necessarily true for other constituents found in the wastewater.[1,2,16,22] It is generally assumed that the inability to grow good turfgrass in soils containing high levels of Na results not from the direct toxicity of Na to the plant, but instead from the negative effect of Na on the soil. Because of these effects on soil quality, especially permeability, an irrigation water is classified according to its Na hazard, expressed as either the sodium absorption ratio (SAR) or adjusted sodium absorption ratio (adj. SAR)[3]. These values give some general indication of the long–term effects of an irrigation water source on soil conditions and crop production. The typical range of Na in wastewater is 50 to 250 mg L^{-1}, and 4.5 to 7.9 for SAR values.[11] Water with values above 70 mg Na L^{-1} would warrant concern when applied to Na sensitive crops. Most established turfgrasses are Na tolerant. Sodium absorption ratio (SAR) values of 3 to 9 would suggest a slight to moderate restriction on water use.

Irrigation water values above 70 mg Na L^{-1} would warrant concern when applied to Na sensitive crops. Most established turfgrasses are Na tolerant.

Grasses such as bermudagrass, crested and fairway wheatgrass (*Agropyron sp.*), tall fescue (*Festuca arundinaceae* Schreb.), annual ryegrass (*Lolium multiflorum* Lam.), alkaligrass (*Puccinellia* spp.), saltgrass (*Distichlis stricta*),

Paspalum sp., and red fescue (*Festuca rubra* L. spp. *rubra*) can tolerate sodic soil conditions in which Na occupies at least 15 percent of the cation exchange sites[3]. However, if these tests had not been performed under stabilized soil conditions, Na–induced poor soil quality probably would have inhibited the growth of even the most Na–tolerant plants.

Sodium accumulation in soils may be controlled through the use of soil amendments such as gypsum (calcium sulfate) and elemental sulfur (S).

Sodium accumulation in soils may be controlled through the use of soil amendments such as gypsum (calcium sulfate) and elemental sulfur (S). These materials are inexpensive and can be applied in a finely ground form that can be watered into the soil. The amount of material to be applied to the soil depends upon how much Na is on the cation exchange sites. It is desirable to reduce the exchangeable sodium percentage (ESP) to less than 10. A useful table for determining the quantity of gypsum or sulfur necessary to reduce the ESP in a 0.3 m soil profile is presented by the U.S. Salinity Laboratory Staff[33].

Mancino and Kopec[21] found the ESP of a wastewater-irrigated turf soil to fall from 10 to 5.1 after a single application of gypsum at 4,480 or 8,960 kg ha^{-1}. This was equivalent to a 250 mg kg^{-1} reduction in Na for this sandy loam soil. In contrast, it required two annual applications of gypsum at 2,240 kg ha^{-1} to reduce soil ESP to 7.4. From a practical standpoint it is not possible to use the highest application rate used in this study because the turf is covered too thickly.

Boron

Boron toxicity does not seem to pose a problem for turfgrass.

The use of various household and commerical detergents results in higher concentrations of B in wastewater than in potable water. Typically, B levels range from <0.1 to 2.5 mg L^{-1}.[24] Levels above 2.0 mg L^{-1} of B require the use of crops listed as moderately tolerant, while levels above 4 mg L^{-1} of B require tolerant and very tolerant crops.[11] However, boron toxicity does not seem to pose a problem for turfgrass. This has been attributed to the fact that B, which accumulates in leaf tips, is removed by frequent mowing.[23]

At the Calistoga Golf Course in Calistoga, California, a high B (4 mg L^{-1}) sewage effluent water was used for irrigation due to drought conditions[9]. Water containing this level of B should only be applied to tolerant (4.0 to 6.0 mg B L^{-1}) or very tolerant (6.0 to 15 mg L^{-1}) crops. No deterioration in putting green quality was reported, presumably due to the practice of clipping removal.

Chlorides

Chloride toxicity is the most common toxicity occurring from irrigation water because it moves readily in the transpirational stream of plants[3]. Toxicity can also occur when chlorides are absorbed directly through the leaves. Crop sensitivity may occur when Cl levels exceed 5 meq L^{-1} in soil solution and 3.3 meq L^{-1} in irrigation water. Turfgrasses, however, have been classified as being relatively Cl tolerant even though grasses readily accumulate Cl.[7,8]

Turfgrasses have been classified as being relatively Cl tolerant even though grasses readily accumulate Cl.

Harivandi[14] reported that some effluent irrigation waters in California contained from 2.8 to 5.2 meq Cl L^{-1}. Current studies by Nelson, Mancino, Pepper and Kopec (data unpublished) using wastewater containing up to 14.3 meq Cl L^{-1} to irrigate turfgrass in Chandler, Arizona have found no apparent turf injury (chlorosis) despite Cl levels in plant tissue that approach 1.5 percent on a dry weight basis. Tissue levels of approximately 0.5 percent are considered normal in non–chloride stressed turf.[8] Cordukes[7] reported slight to moderate injury in various cool–season grasses containing from 1.0 to 2.0 percent Cl in leaf tissue.

Other Trace Elements

Page and Chang[25] discuss the soil and groundwater fate of trace elements found in wastewaters. They point out that these elements can always be found in domestic wastewater. However, they are found at higher levels if the processing plant handles industrial waste discharge. Most of these elements are removed from the wastewater when adsorbed or precipitated by suspended solids and are generally removed during wastewater treatment.

Trace elements are found at higher levels if the processing plant handles any industrial waste discharge.

181

Because of this, sewage sludges may be high in potentially hazardous trace elements.

Most wastewaters meet the criteria set forth by the U.S. Environmental Protection Agency[32] for irrigation water quality. Page and Chang[24] point out that wastewaters which do not meet these criteria may still be suitable for irrigation if they are used carefully. Dudeck[10] presented heavy metal data for wastewater utilized for turf irrigation in Florida. Concentrations of cadmium, copper, nickel, lead, and zinc were below recommended drinking water standards. Hayes et al.[16] and Mancino and Pepper[22] showed that the accumulation of Fe, Cu, Mn, and Zn in effluent-irrigated turf soil did not occur beyond the normal range for agricultural soils[20].

Leaching Losses

Nitrogen

Nitrogen loss via leaching is a primary concern with the use of N–containing wastewater for irrigation because it is assumed that water movement beyond the turf root system will occur. Many studies have examined this type of loss from turf irrigated with wastewater. Sidle and Johnson[30] determined the capacity of turfgrass soil systems to purify municipal sewage effluent and the ability of this system to utilize N from the effluent. The effluent was secondarily treated and chlorinated. It contained from 13 to 26 mg N L^{-1}, primarily in the ammonium (NH_4^+) form. Greenhouse studies were conducted on common bermudagrass grown in either a sandy loam, or silt loam soil, and overseeded with annual ryegrass. Wastewater at rates of 3.3, 6.5, and 7.2 cm per week were applied whenever the soil water content reached 60 percent of field capacity. Ammonium concentrations in the leachate were higher (1.38 mg L^{-1}) from the silt loam soil than from the sandy loam soil (0.36 mg L^{-1}). Nitrate concentrations also were higher in the silt loam soil receiving the highest irrigation regime, but never exceeded 4.0 mg L^{-1}. Overall the turf was able to remove at least 90 percent

Overall the turf was able to remove at least 90 percent of the N delivered by the wastewater even when water was applied at rates resulting in a 41 percent leaching fraction.

of the N delivered by the wastewater, even when water was applied at rates resulting in a 41 percent leaching fraction. The maximum permissible level of nitrate allowed in drinking water is 45 mg L^{-1} [32].

Anderson et al.[1] tried to determine the maximum loading rates of secondary wastewater to a turf–soil system without resulting in leachate having greater than 10 mg N L^{-1} (45 mg NO_3^- L^{-1}). They utilized field lysimeter units measuring 1 m^2 x 60 cm deep. Ten units contained sand (95 percent sand, 1 percent silt, and 4 percent clay) while the remainder contained a soil mixture (89 percent sand, 5 percent silt, 4 percent clay, and 2 percent organic matter). Both soils had a pH of 8.3 and were seeded to common bermudagrass.

Nitrogen lost in leachate was primarily in the NO_3^- form and it increased as irrigation rates increased.

Units were overseeded to annual ryegrass during the winter months. The lysimeters were drip irrigated with wastewater, and leachate was collected for analysis. Irrigation rates were equivalent to 7.0, 11.6, 15.2, 24.0, and 30.4 cm wk^{-1}. One–half of each rate was applied twice a week. The wastewater contained 17 to 23 mg NH_4^+-N L^{-1} and 0.5 mg NO_3^- N L^{-1} during the summer months, and 3 to 6 mg NH_4-N L^{-1} and 4 to 16 mg NO_3^--N L^{-1} in the winter.

Nitrogen lost in leachate was primarily in the NO_3^- form and it increased as irrigation rates increased. Leachate concentrations of NO_3^--N did not exceed 10 mg L^{-1} in the sand soil until wastewater was applied at 11.6 cm wk^{-1} and 15.2 cm wk^{-1} for the soil mix. Maximum NO_3^--N concentrations were 17.7 and 14.2 mg L^{-1} from the sand and soil mix, respectively, when the irrigation rate was 30.4 cm wk^{-1} at the lowest irrigation level. During summer months a peak irrigation rate of only 5 to 6 cm would be expected on southern Arizona golf courses. Ammonium-N in leachate was less than 1 mg L^{-1}.

NO_3^--N concentrations in leachate increased during December, January, February, June and July when NO_3^- levels in waste-water were highest and turf growth was slowest.

In similar work, Pepper et al.[25] found N levels in leachate increased as wastewater application rate increased when applied to sand:silt:clay soil mixes having cation exchange capacities of 3.5, 4.9, or 7.7 meq $100g^{-1}$ soil. Leachate values rarely exceeded 10 mg NO_3^--N L^{-1} for any of the soil mixes regardless of very high wastewater application rates (up to 36.4 cm wk^{-1} applied in two equal applications per week). They found NO$_3$-N concentrations increased during December, January, February, June and July when NO$_3$ levels in wastewater were highest and turf growth was slowest. They speculated that high rates of evapotranspiration in the summer resulted in a concentration of NO$_3$-N in the leachate.

More recently Hayes et al.[16] found that following 16 months of wastewater irrigation, leachate from bermudagrass turf plots had NO$_3$-N levels averaging 22.3 mg kg^{-1}, or 7.8 mg kg^{-1} greater than turf plots receiving potable irrigation plus 48.4 kg urea-N ha^{-1} mo^{-1}. The wastewater used contained approximately 49 kg N in every 0.12 ha-m applied, predominately in the NH$_4^+$-N form and annual application rates were limited to 0.62 ha-m yr^{-1}. This study suggests that a potential does exist for NO$_3$-N leaching losses, so the turf manager should irrigate in a manner to minimize leaching.

Phosphorous

Wastewaters typically contain about 10 mg P L^{-1} while the typical range of irrigation water is 0 to 2 mg L^{-1}.

Phosphorous leaching or runoff can also result in the contamination of ground and surface waters. Wastewaters typically contain about 10 mg P L^{-1} while the typical range of irrigation water is 0 to 2 mg L^{-1}.[6,3] Wastewater in Tucson, Arizona actually delivers P at rates similar to those of N (398 kg ha-m water) and ranging from 6.4 to 26.8 mg L^{-1}.[16] The precipitation of P with Fe, aluminum (Al), Ca and Mg in soils generally reduces its potential loss through leaching. However, leaching depends upon the P loading rate, quantity of irrigation water applied, soil texture and soil depth. Many turfgrass sites are located on sandier soil where P fixation could be low.

Pepper et al.[25] and Anderson et al.[2] reported that P leaching would not be the limiting factor in wastewater irrigation rates if the turf–soil system was to be used to further purify wastewater for the purpose of groundwater recharge. Lechate concentrations of P rarely exceeded 3 mg L^{-1} in the summer and 5.4 mg L^{-1} in the winter from their sandy soil (CEC = 3.5 meq 100g) even though the wastewater averaged 6.8 mg P L^{-1} and was applied at rates up to 30.4 cm wk^{-1}.

In an effluent–irrigated sandy loam turf soil, Hayes et al.[16] reported P levels to increase by about 20 mg kg^{-1} soil over the original P levels in the unirrigated soil (about 20 mg kg^{-1} soil). On the other hand, P levels decreased in potable–irrigated plots, presumably due to plant uptake. Mancino and Pepper[22], continuing the work of Hayes et al.[16,17], found P levels to increase to 35 mg kg^{-1} after 3.5 years. These P levels should be considered more than adequate to meet the needs of the turfgrass and additional P would not be required from fertilizer. Phosphorous levels decreased to 10 mg kg^{-1} soil when irrigation was temporarily halted.

In some research, P levels were more than adequate to meet the needs of the turfgrass and no additional P need be applied from fertilizer.

Phosphorous can result in the precipitation of other plant available nutrients in the soil such as Ca, Mg, and Fe. Hayes et al.[17] reported a greater incidence of Fe chlorosis on effluent irrigated turf, possibly resulting from P combining with Fe to produce insoluble Fe phosphates[13].

Phosphorous can result in the precipitation of other plant available nutrients in the soil such as Ca, Mg, and Fe.

Boron and Chlorides

Boron is not retained by the soil because it is an anion, but boron compounds (borate) are moderately held by soil cation exchange sites. Some potential for B leaching exists, especially in sandy soils. Anderson et al.[2] reported effluent–irrigated soil leachate concentrations of B to be equal to that of the concentrations found in the wastewater (0.75 mg L^{-1}). This would indicate a continuous leaching of B through the soil profile.

Some potential for boron leaching exists, especially in sandy soils.

Pepper et al.[25] also reported that B levels in leachate from effluent-irrigated turf were similar to that of the wastewater (0.25 to 0.71 mg B L^{-1}). Mancino and Pepper also found boron levels to remain low (<1 mg kg^{-1} dry soil) in effluent–irrigated turf soil, remaining similar to that of the wastewater used to irrigate it when a leaching fraction of about 20 percent is used.

Chlorides leach easily through the soil because they are anionic. Because of this, Cl is often used in soil to model

Little information is available on Cl leaching from waste-water–irrigated turf.

the leaching loss of NO_3^-. Little information is available on Cl leaching from wastewater–irrigated turf. Anderson et al.[2] reported that Cl could not be found in leachate from turf growing on either a sand or sand:soil mix. Chloride levels in the wastewater were low (0.4 mg L^{-1}) because of sunlight–induced degradation during storage in the wastewater lagoons. Soil Cl levels in a current study in Chandler, Arizona are about 550 mg kg^{-1} soil in the top meter of soil. These concentrations nearly match that of the wastewater (360 to 550 mg L^{-1}). This Cl may be available for leaching.

Human Health Concerns

The likelihood of human disease occurring through the use of secondary or tertiary treated wastewater for turf irrigation is low. However, because a large number of enteric viruses and parasites are found in raw sewage, a potential does exist for disease causing organisms to survive the wastewater treatment process and become reintroduced into environments where direct human contact may occur.

The likelihood of human disease occurring through the use of secondary or tertiary treated wastewater for turf irrigation is low.

Frankenberger,[12] in his discussion of the fate of pathogens from wastewater in soil and groundwater, shows that there is considerable evidence that human pathogens can survive the wastewater treatment process. It has been reported that *Giardia* and *Cryptosporidium* cysts and oocysts can be found in secondary treated wastewater.[27,28,31] Rose and Gerba[29] found 14 percent of secondary or post–secondary treated wastewater from treatment facilities in Arizona and Florida to be positive for enteroviruses. Thirteen percent were in violation of unrestricted irrigation access standards (1 plaque forming unit/40 L) for Arizona, but none were in violation of restricted access standards (125 plaque forming units/ 40 L). Forty-one percent of all wastewater samples contained *Giardia* cysts and 33 percent were in violation of Arizona's health standard (1 cyst/40 L) despite post-secondary wastewater treatment. Howard[18] reported finding live human parasitic worms in sprinkler head filters at a private golf course in Phoenix, AZ. These worms were most likely an intestinal worm (*Ascaris lumbricoides*).

Whether or not human pathogens can survive in or pass through a turf–soil system has been examined. Pepper et al.[25] determined whether different human pathological viruses (after seeding into non–chlorinated secondary effluent) and *E. coli* (an indicator organism) could pass through various turfgrass soils (all predominately sand) at high wastewater loading rates (up to 36.4 cm/week). Despite repeated pathogen seeding the turf–soil system was highly efficient (99 percent or greater) in removing poliovirus type 1, echovirus type 5, and Coxsackie B3 virus and *E. coli*. It was reported that higher soil CEC capacities resulted in greater virus removal.

Despite repeated pathogen seeding the turf–soil system was highly efficient (99 percent or greater) in removing poliovirus type 1, echovirus type 5, and Coxsackie B3 virus and E. coli.

Badaway et al.[4] decided to determine the survival of enteric viruses and coliphage on turf irrigated with unchlorinated secondary effluent because of documented pathogen survival on wastewater irrigated crops and vegetables, and in soil. The turf (hybrid bermudagrass in the summer and annual ryegrass in the winter) was grown outdoors in full sunlight in 20 x 60 cm wooden planters. Poliovirus type 1 and rotavirus SA-11 had a 99.8 percent reduction after 10 hours of exposure to the outdoors on bermudagrass leaves and no polio virus could be detected after 11 hours. Disappearance was rapid once temperatures increased above 38°C. During the winter (4 to 16°C) coliphage had a 99.99 percent reduction after 24 hours and a 99.99999 percent (as reported) reduction after 40 hours. Poliovirus decreased by 96 percent and 99.6 percent after 24 and 40 hours, respectively. Rotavirus could not be recovered after 40 hours. Most of the disappearance occurred between 6:30 and 8:30 a.m. when air temperature and light intensity increased.

Wastewater irrigation may represent a potential small risk of disease, particularly when used on play-grounds and parks where children may roll in the grass or even ingest grass.

Badaway et al.[4] concluded that wastewater irrigation "may represent a potential small risk of disease, particularly when used on playgrounds and parks where children may roll in the grass or even ingest grass." These authors also suggested that standards for viruses in wastewater to be used on unrestricted access landscape areas, such as public parks and playgrounds, be made to match the standard for food crops (1 plaque forming unit/40 L).

Howard[18] recommended that turf managers monitor the residual chlorine levels in effluent–containing reservoirs to insure adequate disinfection.

Summary

Based on scientific data and the past experience of many turfgrass managers, it appears that secondary or post-secondary treated wastewater can be used successfully to irrigate turfgrasses. Most data indicate that the benefits of using wastewater exceed those of the detriments. However, as with any irrigation water, its suitability must be assessed through careful water and soil analyses. Waters with high salinities, Na, B, Cl, or other trace elements should be avoided if possible. When the use of this type of water is mandated, consideration should be given to the mixing of the lower quality water with a higher quality water prior to application. The turfgrass manager should also take into account additional plant nutrients being provided by the wastewater to minimize the potential for nutrient leaching and to promote healthy turf. Besides turf, the effects of wastewater irrigation on other landscape plants must also be taken into consideration.

Most data indicate that the benefits of using wastewater exceed those of the detriments.

References

1. Anderson, E.L., I.L. Pepper, and W.R. Kneebone. 1981a. Reclamation of wastewater with a soil-turf filter: I. Removal of nitrogen. J. Water Poll. Control Fed. 53:1402-1407.

2. Anderson, E.L., I.L. Pepper, W.R. Kneebone, and R.J. Drake. 1981b. Reclamation of wastewater with a soil-turf filter:II. Removal of phosphorous, boron, sodium, and chlorine. J. Water Poll. Control Fed. 53:1408-1412.

3. Ayers, R.S., and D.W. Westcott. 1985. Water quality for agriculture. FAO Irrigation and Drainage Paper #29, Rome, Italy.

4. Badaway, A.S., J.B. Rose, and C.P. Gerba. 1990. Comparative survival of enteric viruses

and coliphage on sewage irrigated grass. J. Environ. Sci. Health 25:937-952.

5. Beard, J.B. 1973. Turfgrass: Science and culture. Prentice-Hall, Englewood Cliffs, NJ.

6. Bouwer, H., and R.L. Chaney. 1974. Land treatment of wastewater. Adv. Agron. 26:133.

7. Cordukes, W.E. 1970. Turfgrass tolerance to road salt. The Golf Superintendent 38:44-48.

8. Cordukes, W.E., and E.V. Parupes. 1971. Chloride uptake by turfgrasses. Can. J. Plant Sci. 51:485-490.

9. Donaldson, D.R., R.S. Ayers, and K.Y. Kaita. 1979. Use of high boron sewage effluent on golf greens. California Turfgrass Culture 29:1-2.

10. Dudeck, A.E. 1978. Current review of sewage effluent for irrigation use. USGA Green Section Record July/August:5-9.

11. Feigin, A., I. Ravina, and J. Shalhevet. 1991. Irrigation with treated sewage effluent: Management for environmental protection. Advanced Series in Agricultural Sciences 17. Springer-Verlag, Berlin, Germany.

12. Frankenberger, W.T. 1985. Fate of wastewater constituents in soil and groundwater: Pathogens. p. 14(1)-14(25). *In* G. Stuart Pettygrove and T. Asano (ed.) Irrigation with reclaimed municipal wastewater—a guidance manual. Lewis Publishers, Inc., Chelsea, MI.

13. Fuller, W.H. 1975. Management of southwestern desert soils. Univ. of Arizona Press, Tucson, AZ.

14. Harivandi, A. 1991. Effluent water for turfgrass irrigation. University of California Cooperative Extension Leaflet 21500.

15. Harivandi M.A., J.D. Butler, and L. Wu. 1992. Salinity and turfgrass culture. *In* D.V. Waddington, R.N. Carrow, and R.C. Shearman (ed.) Turfgrass. Agronomy 32:207-229.

16. Hayes, A.R., C.F. Mancino, and I.L. Pepper. 1990. Irrigation of turfgrass with secondary sewage effluent: I. Soil and leachate water quality. Agron J. 82:939-943.

17. Hayes, A.R., C.F. Mancino, W.Y. Forden, D.M. Kopec, and I.L. Pepper. 1990. Irrigation of turfgrass with secondary sewage effluent: II. Turf quality. Agron J. 82:943-946.

18. Howard, H.F. 1992. Irrigating with effluent. Grounds Maintenance March:52-58.

19. Kneebone, W.R., and I.L. Pepper. 1984. Luxury water use by bermudagrass. Agron. J. 76:999-1002.

20. Lindsay, W.L., and W.A. Norvell. 1978. Development of a DTPA soil test for zinc, iron, manganese, and copper. Soil Sci. Soc. Am. J. 42:421-428.

21. Mancino, C.F., and D.M. Kopec. 1989. Effects of gypsum on a wastewater irrigated turfgrass soil. p. 39-46. *In* D.M. Kopec (ed.) University of Arizona Turfgrass and Ornamental Research Summary Series P-80.

22. Mancino, C.F., and I.L. Pepper. 1992. Irrigation of turfgrass with secondary sewage effluent: Soil quality. Agron. J. 84:650-654.

23. Oertli, J.J., O.R. Lunt, and V.B. Youngner. 1961. Boron toxicity in several turfgrass species. Agron. J. 53:262-265.

24. Page, A.L., and A.C. Chang. 1985. Fate of wastewater constituents in soil and groundwater: trace elements. p. 13(1)-13(16). *In* G. Stuart Pettygrove and T. Asano (ed.) Irrigation with reclaimed municipal wastewater—a guidance manual. Lewis Publishers, Inc., Chelsea, MI.

25. Pepper I.L., W.R. Kneebone, and P.R. Ludovici. 1981. Water reclamation by the use of soil-turfgrass systems in the southwest USA. U.S. Dept. of Interior Ofc. Water Res. and Tech. B-072-ARIZ.

26. Pettygrove, G.S., and T. Asano. 1985. Irrigation with reclaimed municipal wastewater—a guidance manual. Lewis Publishers, Inc., Chelsea, MI.

27. Rose, J.B. 1986. Microbial aspects of wastewater reuse for irrigation. CRC Critical Reviews in Environmental Control 16:231-256.

28. Rose, J.B. 1988. Occurrence and significance of *Crytosporidium* in water. J. Amer. Water Works Assoc. 80:53-58.

29. Rose, J.B., and C.P. Gerba. 1990. Assessing potential health risks from viruses and parasites in reclaimed water in Arizona and Florida, USA. Wat. Sci. Tech. 23:2091-2098.

30. Sidle, R.C., and G.V. Johnson. 1972. Evaluation of a turfgrass–soil system to utilize and purify municipal wastewater. Hydrology and Water Resources in Arizona and the Southwest 2:277-289.

31. Sykora, J.L., S.J. States, W.D. Bancroft, S.N. Boutrous, M.A. Shapino, and L.F. Conley. 1986. Monitoring of water and wastewater for *Giardia*. p. 1043-1054. *In* Advances in Water Analysis and Treatment Water Quality Technology Conference. Amer. Wastewater Assoc., Denver, CO.

32. U.S. Environmental Protection Agency. 1973. Water quality criteria. Ecological Research Series, EPA R3-73-033, U.S. EPA, Washington, DC.

33. U.S. Salinity Laboratory Staff. 1954. Diagnosis and improvement of saline land alkali soils. USDA Handbook No. 60. U.S. Gov. Print Office, Washington, DC.

Coping with Wastewater:
Management Strategies and Information Sources

James Moore
Director
USGA Green Section
Mid–Continent Region

Introduction

Non–golfers often fail to see any significant redeeming value to the existence or maintenance of a golf course.

Non–golfers often fail to see any significant redeeming value to the existence of a golf course, and many seem to feel that the maintenance of a golf course damages the community environment. Specifically, irrigation of the course and application of fertilizers and pesticides often is perceived as a cause of groundwater contamination and groundwater depletion. The USGA is presently funding research addressing the fate of fertilizers and pesticides in the soil. This research is verifying that the combination of turfgrass and soil serves as a very effective filtering and cleansing system. With proper management of the various chemicals applied to a course, groundwater contamination should not occur. However, the irrigation of a golf course does require the use of large quantities of water.

All communities must deal with wastewater. After treatment, many simply discharge this water into the nearest stream or river, where it eventually reenters the hydrologic cycle. While properly treated wastewater does not represent a threat to the potable water supply, direct discharge back into that supply is far less effective in terms of filtering than application to a turf/soil system.

Golf courses need water and communities would benefit from a better place to discharge effluent than into their streams and lakes.

Golf courses need water, and communities would benefit from a better place to discharge effluent than into their streams and lakes. It seems to be the perfect combination. Other speakers at this symposium have detailed the science and research that helped identify the steps necessary to be certain this "marriage" is a successful one. However,

192

science is only half the challenge. The other half is convincing people to accept this union.

For quite some time after being asked to participate in this meeting, I struggled with how I could make some sort of real contribution. My qualifications are not those of a scientist, but rather those of an agronomic practitioner. My strength is experience; experience gained from growing up on a golf course, working on courses part–time as much as possible during my military service, working as a golf course superintendent, and finally, experience gained over the past eight years on the USGA staff.

What I plan on sharing with you today are the most often voiced concerns of players and superintendents regarding wastewater on the golf course. I will discuss with you the suggestions I offer to superintendents concerning course management as it relates to wastewater, how I try to convince golfers to utilize this resource, and what sources of information I use to justify my opinions and try to dispel their fears.

Member/Player Concerns

Unfortunately, it has been my experience that golfers will not support the use of effluent water until it is forced upon them. The connotations associated with the words "waste," "effluent," or "gray" water all tend to inspire fear rather than promote thoughts of environmentalism, conservation, or fiscal responsibility. It is for this reason that I strongly agree with the other speakers—this resource should not be referred to as "wastewater" but rather as "recycled water."

The words "waste," "effluent," or "gray" water all tend to inspire fear rather than promote thoughts of environmentalism, conservation, or fiscal responsibility.

With increasing frequency, golfers and golf course superintendents across the country are starting to consider the use of recycled water. Understandably, they have many questions. While they may be presented with volumes of research documenting the scientific aspects

193

of recycled water, invariably the same questions eventually will be asked. These questions are:

Will it smell?

Possibly. But as opposed to what? Most water used to irrigate golf courses has a smell. I have vivid memories of a small sod farm I managed while in school. There was so much natural gas in the well water that I could set fire to the vapors. While this was a great trick to play on visitors, the real trick was drinking the water due to the odor. As someone who for much of the year spends each night in a different city, I can assure you that I notice distinct odors in the potable water supply from one community to the next. Storage lakes on golf courses (complete with fish, ducks, algae, etc.) often lend a strong odor to the irrigation water. The point is, almost all water used to irrigate a golf course has some odor; it's just a matter of getting used to it. The odor is usually strongest early in the irrigation cycle, perhaps due to small accumulations of gases in the air pockets of the irrigation system. Since irrigation when golfers are present is avoided as much as possible regardless of the water source, most players will not notice the odor.

I have visited courses in other countries where untreated sewage was the source of the irrigation water. For many reasons, raw sewage should never be used, not the least of which is a smell that at first can knock you over. However, even under such extreme conditions, you quickly grow accustomed to the odor.

Yes, there is a chance wastewater will have an odor the player might notice. However, the odor will not be one of sewage, but rather of the chemcials used in its treatment.

Will it detract from the appearance of the course?

No. The key here is proper storage. Wastewater is no more unattractive than any other water source when properly stored. Conversely, any source of water will

Questions eventually asked by golfers and surrounding homeowners:

- *Will it smell?*
- *Will it detract from the appearance of the course?*
- *Will we be liable?*
- *Will it affect our property value?*
- *Will it make us sick?*
- *Will it stain our houses?*

Wastewater is no more unattractive than any other water source when properly stored.

detract from the appearance of a facility when stored in a reservoir that is improperly constructed or managed. Aesthetically it is true that *improperly* stored wastewater is more likely to cause aesthetic problems than other sources of water. For example, nutrients in wastewater encourages the bloom of algae, making its control more difficult and the pond less aesthetically appealing . The major reservoir problems are in lakes that are much too shallow, inadequately aerated, choked with algae, and contaminated with silt and other debris as a result of runoff.

The major reservoir problems are lakes that are much too shallow, inadequately aerated, choked with algae, and contaminated with silt and other debris as a result of runoff.

Our industry is fortunate that in recent years many new tools and techniques have been introduced to improve reservoir management. For lakes that are too shallow we now have efficient dredging equipment and affordable lake liners. The grass carp is a wonderfully efficient lake cleaner that is being legalized in more and more states. New aeration equipment (both subsurface and surface) makes the lake much more attractive and improves the overall quality of the water. Concerns about pesticide and fertilizer runoff have resulted in the planting of buffer strips of native vegetation which very effectively reduce silt and other debris from entering the reservoir. Finally, today's contractors are more aware of the need to provide additional depth to reservoirs. This new awareness has been inspired by increasing concern over the amount of water used by golf courses which in turn promotes designs with greater storage capacity.

Will we be liable?

Given the current state of our society, it seems that no matter what you do you will be held liable for something sooner or later. The only liability problems I have encountered over the past eight years have arisen from poor treatment of the effluent. This is a failure in the operation of the treatment facility and a failure in the monitoring program of the end user. Courses built with inadequate regard for surface drainage and poorly designed irrigation systems are most vulnerable to this liability. Superintendents who overwater (for whatever reason) also are easy targets.

The only liability problems I have encountered over the past eight years have arisen from poor treatment of the effluent.

Will it affect our property values?

Many courses in high profile and wealthy neighborhoods utilize wastewater.

This question periodically arises on courses that are forced to switch to wastewater. I am unaware of any research to document the reduction in property values. To my knowledge it has not occurred. In addition, many new amenity courses utilize effluent water from the development's own sewage treatment station. As a result, many courses in high profile and wealthy neighborhoods utilize wastewater. Property values do not appear to have suffered in this area as a result of wastewater use.

Will it make us, our children, or our pets sick?

The key is good monitoring of the water supplied from the treatment plant and an irrigation system that is designed to allow virtually all irrigation to take place at night.

Fortunately, there are a number of good references that detail the levels of treatment necessary to prevent illness from exposure to recycled water. From a turf management view, the key is good monitoring of the water supplied from the treatment plant and an irrigation system that is designed to allow virtually all irrigation to take place at night. Another very important point here is the perception of risk. It is arguable that any actual threat from illness might occur. Again, it seems our major task is to educate those whose unfamiliarity with the composition of recycled water leads to fear of its use.

Will it stain our houses?

I have yet to hear any occurrence of staining of sidewalks, houses, or cars with effluent water.

When I was asked this in a general membership meeting, I almost laughed out loud. Unfortunately, no one else in the meeting was laughing. This illustrates the fear factor associated with anything referred to as "waste." To address the question, I have yet to hear of any occurrence of staining of sidewalks, houses, or cars with effluent water, although I should point out the potential for staining when domestic water contains high levels of iron.

Superintendent Concerns

Given the agronomic experience and education of most of today's golf course superintendents, it is not surprising that these folks tend to ask much more difficult questions.

Is it of good quality?

The information to evaluate water quality is available from a number of sources. The real trick is evaluating all the related factors.

Are the greens built well enough to avoid problems with recycled water?

Greens vary widely in their construction. Often, recycled water will be higher in salts than potable water from the same area. Surface and subsurface drainage, and infiltration rates are the most important green construction considerations when evaluating any source of water.

Many older greens use a root zone mix composed primarily of loam, clay loam and clay soil. Such greens have low infiltration rates (often less than 1 inch per hour) and are almost totally dependent on surface drainage (the runoff of excess water from the putting surface). As cutting heights on greens have dropped over the years, these older, more slowly permeable greens have been placed under a great deal of extra pressure. A large number are already maintained on the edge of failure, thus requiring extremely close management on the part of the superintendent. Switching to any water source that is less favorable in terms of quality will place even greater stress on such greens and will greatly increase the chance of turf failure.

New aerification tools have improved the internal drainage of many older, soil–based greens. However, most greens with this type of construction do not have tile drainage systems. While aerification can help move salts deeper into the profile, it seldom provides a long–term solution. Reconstruction to provide a more rapidly draining

Superintendent concerns:
- *Is it good quality water?*
- *Are the greens built well enough to avoid problems with wastewater?*
- *Are the nutrient levels consistent?*
- *Can the irrigation system handle wastewater?*
- *How much water will they get?*

Surface and subsurface drainage, and infiltration rates are the most important green construction considerations when evaluating any source of water.

root zone, a subsurface drain tile system, and improved surface drainage, may well be necessary when using recycled water that is high in salts.

Are nutrient levels consistent?

Today's golf course superintendent closely manages nutrient levels throughout the course. Most superintendents know exactly within 1/2 lb./1000 sq. ft. how much of each element has been applied to the greens. Often, the agronomic procedures necessary to produce a healthy turfgrass conflict with the procedures necessary to produce a putting surface that is acceptable to the players. One of the most influential maintenance practices affecting the speed of a putting green is nitrogen fertilization. Increased growth rate of the turf results in slower putting speeds. Not too many years ago, it was common practice to apply one pound or more of nitrogen per thousand square feet in a single application. The turf responded with vigorous growth and the speed of the greens decreased accordingly. Today's superintendents strive for consistent green speeds. Nitrogen is now most often applied in tenths of a pound—a practice commonly referred to as "spoon feeding."

Superintendents rightly fear a loss of control over nutrients applied to the greens when recycled water that is high in nutrients is used as an irrigation source.

Superintendents rightly fear a loss of control over nutrients applied to the greens when recycled water that is high in nutrients is used as an irrigation source. Since the amount of nitrogen applied through irrigation with recycled water is relatively small, superintendents can adjust their fertilizer program accordingly—*assuming the level of nutrients in the recycled water remains consistent.* Water that varies widely in the percentage of nitrogen, phosphorous, and other elements will reduce the superintendent's control over fertility.

Close attention also should be given to the amount of nitrogen applied by virtue of irrigation each month.

Close attention also should be given to the amount of nitrogen applied by virtue of irrigation each month. The timing of nitrogen fertilization is critical on bentgrass greens in the summer and on bermudagrass greens during spring transition. A superintendent therefore would be wise to request a historical analysis of the recycled water to determine consistency of the nutrient content through-out the year.

Can the irrigation system handle it?

Given the concerns over health and smell, it is advan-tageous to apply all recycled water during times golfers are not on the course. Many irrigation systems simply are not designed to apply all the water the course requires during nighttime hours. This is particularly true on older courses where the original design has been altered over the years as pipe and heads are extended into previously unirrigated areas. Limiting factors are most often the capacity of the pump station, size of the pipe, and degree of control provided by the automatic system.

Given the concerns over health and smell, it is obviously advan-tageous to apply all recycled water during times golfers are not on the course.

Fortunately, the technology of pump station and irrigation control (both central and satellite) has improved tremen-dously over the past few years. Computerized systems now give the superintendent precise control over which heads come on, for how long, and in what order. Variable speed pump control ensures peak efficiency of the pump station in spite of frequent changes in irrigation sched-uling. As a result, the operation of the irrigation system can be quickly and accurately adjusted. Unfortunately, pipe size cannot be so easily modified. It is quite possible that the older irrigation system will have to be completely replaced when the switch to recycled water is made.

Limiting factors are most often the capac-ity of the pump station, size of the pipe, and degree of control provided by the automatic system.

How much water will we get?

This is one of the first questions the golf course super-intendent should ask. Every good turf manager recog-nizes the importance of good water management. Overwatering is arguably the most serious "sin" in turfgrass management. The health of the turf and the playing quality of the course both suffer severely when too much water is applied for an extended period of time. It therefore is imperative that the golf course not be forced to accept more water than it needs. Adequate storage must be provided or arrangements made to divert excess water to another use; or to store it, or arrange for it to be carried away by a stream or injected into a deep aquifer.

Overwatering is arguably the most serious "sin" in turfgrass management. It therefore is impera-tive that the golf course not be forced to accept more water than it needs.

No course should enter into an agreement that allows irrigation to be controlled by the sewage treatment plant rather than the golf course superintendent.

Unfortunately, there have been instances where such foresight was not practiced and an agreement was made that required the course to accept a fixed quantity of water *regardless of climatic conditions*. No course should enter into an agreement that allows the single most important and influential maintenance practice (irrigation) to be controlled by the sewage treatment plant rather than the golf course superintendent.

Sources of Information

Provide scientific information supporting the use of this resource.

A critical component of the "selling" process (convincing people that the use of recycled water is desirable) is to be able to provide scientific information supporting the use of this resource. Over the years as an agronomic consultant, I have found the following sources to be extremely valuable.

Turfgrass Information File (TGIF)

The TGIF represents the world's largest collection of turfgrass literature organized into a computerized database.

The TGIF represents the world's largest collection of turfgrass literature organized into a computerized database at the Michigan State University Library. The Turfgrass Information File was developed through the support of the USGA as a service to golf. It offers current and historical references on virtually every aspect of turfgrass maintenance. Nearly 200 abstracts concerning recycled water were located and downloaded to my office computer in less than 30 minutes in preparation for this presentation. The same information could have been obtained through the mail for those who do not wish to use a computer and modem to access TGIF. The Turfgrass Information File is available to anyone at nominal fees. For additional information or to utilize TGIF immediately via the telephone, contact Mr. Peter Cookingham at the following number.

Mr. Peter Cookingham
Project Manager, TGIF
Michigan State University Library
W 212 Main Library
East Lansing, MI 48824-1048
517-353-7209

Other books that have proved invaluable in providing technical information concerning the use of recycled water are listed below.

Golf Course Management & Construction:
Environmental Issues

Handbook of Ground Water Contamination,
Assessment, and Remediation

Landscape Restoration Handbook

Irrigation With Reclaimed Municipal Wastewater—
A Guidance Manual

The Water Encyclopedia

All of these texts can be obtained from the publisher listed below:

Lewis Publishers, Inc.
121 S. Main Street, P.O. Drawer 519
Chelsea, MI 48118

Chapter 5 Feasibility of Using Wastewater on Golf Courses

Wastewater Irrigation for Golf Courses: Advantages versus Disadvantages

Charles H. Peacock, Ph.D.
Professor, Dept. of Crop Science
North Carolina State University

Retrofitting a Golf Course for Recycled Water: An Engineer's Perspective

Charles Z. Steinbergs, P.E.

Princeton Meadows Country Club

Paul Powondra

Kiawah Island, South Carolina

George Frye

The Use of Effluent Irrigation Water: A Case Study in the West

Mike Huck

Barton Creek Wastewater Case Study

Tim Long

Wastewater Use for Turf: The Non–Technical Economic Side

William S. Rodie

Wastewater Irrigation for Golf Courses: Advantages versus Disadvantages

Charles H. Peacock, Ph.D.
Department of Crop Science
North Carolina State University

Introduction

Concerns over the use of wastewater for irrigating golf courses must address both agronomic and personal points. One of the most difficult problems is that of image. The term "wastewater" in itself implies something which is unwanted, discarded, or worthless after being used[17]. This connotation projects a very negative image from a public perception. "Wastewater" is a very valuable commodity from an irrigation viewpoint and in many situations it is the difference between having supplemental water for turf culture or having none at all. A better choice in terms is "recycled" or "reclaimed" water, with recycled fitting better into the development of 21st century philosophy regarding conservation of natural resources.

One of the most difficult problems is that of image. The term "wastewater" in itself implies something which is unwanted, discarded, or worthless after being used.

While this chapter will focus primarily on issues related to agronomic concerns, it will include some concerns from an image or personal perspective. Additionally, while it will specifically address some of the problems associated with use of recycled water for irrigation, this does not imply they are not manageable at some level. Where problems occur, it is necessary for the golf course superintendent to develop cultural programs which address these concerns and minimize their impact. All of the issues addressed here will not necessarily occur at each site. An on–going program which includes close scrutiny of water quality, edaphic (soil related) interactions, and plant growth is critical to ensuring that recycled water can be used for irrigation and turf can be grown successfully. Thus, any disadvantage does not mean it cannot be "managed," but simply that dealing with these problems necessitates the superintendent have a sound

Where problems occur, it is necessary for the golf course superintendent to develop cultural programs which address these concerns and minimize the impact if these situations occur.

understanding of the interaction of soils, plant growth and development, and the influence of the environment on management decisions.

Three major categories where management is affected by the use of recycled water include the following:

1. Turf and soils management—the greatest effect is on regulation of growth and prevention of unwanted changes in soil conditions which may adversely affect growth;

2. Irrigation system—the impact on system operation and maintenance since regulatory issues may dictate changes be made in where, when, and how recycled water will be used;

3. Natural phenomena—this includes everything from the weather to health considerations.

Major categories for which turf management is affected:
- *turf and soils*
- *irrigation system*
- *natural phenomena*

Water Conservation

Much of the western United States has experienced extreme drought over the last seven years due to a shift in winter precipitation patterns. This pattern changed in 1993 and brought much needed relief to the area. However, California Governor Pete Wilson commented that despite the fact that there would be an end to water rationing, there would not be an end to water conservation. This must be the attitude developed, not just in California, but nationwide to ensure that potable water supplies are reserved for personal consumption.

Any area developed for recycled water application must take into account certain design considerations[2,5]. These include the following: (1) soil type and topography; (2) characterization of the recycled water, including rate of flow, nutrient loads, organic loads, and concentration of any toxic materials; and (3) assimilative capacity of the site, which is related to soil type, type of vegetation planted, and environmental conditions to which the site is exposed[9].

Any area which is to be developed for recycled water application must take into account certain design considerations:
- *soil type and topography*
- *characterization of wastewater source*
- *assimilative capacity of the site*

Recycled Use Advantages

Use of recycled water on a golf course would be better than surface water or ocean dumping.

Paramount in this category is the fact that there is water available for irrigation use rather than none at all. Additionally, since this supply is dependent on human consumption and supply, it is available based on demand rather than rationed during short– or long–term droughts. Another distinct advantage is the benefit of using a natural filtration system for effluent disposal from sewage treatment facilities. Benefits of this have long been known and this is much more desirable than surface water or ocean dumping or deep well injection[15].

Recycled Use Disadvantages

There may be supply problems, with either a deficiency or excess of recycled water. In resort areas where populations shift seasonally, the sanitary waste stream may fluctuate greatly. If the facility supplying recycled water is a minimum capacity system serving a small seasonal community, recycled water flows could change significantly from month to month. This would mean having to use a supplemental source which could be of either poor quality or great expense.

Conservation disadvantages for recycled water use:
* *deficiency or excess supply*
* *proper site design and turf considerations*
* *potential surface or groundwater contamination*

Another more prevalent problem is whether the site is engineered as a recycled water disposal system[6]. With this situation, over–application easily occurs. The assimilative capacity of the site, including hydraulic loading (the amount of water the site can infiltrate and percolate) is not necessarily a measure of the ideal relationship for optimum irrigation of turf. While turf growth may be manageable under excess irrigation, the impact of traffic and play on these areas may be costly considering revenues and increased maintenance factors. Irrigation amounts and rates should be gauged to turf demands, not site capacity. This includes allowing resting periods from irrigation to allow for reaeration of the soil. Enough flexibility must be engineered into the site design to allow for rainy periods, when no recycled water is used and for an adequate recovery period for drainage and aeration[16]. Typical hydraulic loading rates based on soil grouping are listed in Table 1. However, even the lowest application rate (Soil Group IVa) may not necessarily correlate with turf use, even under semi–arid or arid conditions which at maximum only approach 0.5 inch/day[11]. Low soil oxygen from over–irrigation or poor drainage results in anaerobic production of lactic acid, ethanol and acetaldehyde, all of which are toxic to root cells[1]. This condition also favors the growth of saprophytic fungi, which affect overall turf vigor partly due to ethanol exudates from root systems.

An area which has not been adequately addressed in research or engineering is the potential for surface or ground water contamination from runoff or percolation of recycled water. Some states do regulate where effluent can be applied in relation to potable drinking water supplies, but application of recycled water and the potential for negatively impacting surface and ground water supplies depends on the quality of the effluent and the potential of the turf–soil system for providing a filtration effect. Adequate soil is essential to achieve adequate filtration of the recycled water, and sites having ground water supplies closer to the soil surface are more vulnerable to problems.

Table 1. Hydraulic Loading Rates Commonly Used in Site Design for Different Soil Groupings

Soil Group	Texture	Application Rate inches/day
I	Sandy	1.5 to 2.0
II	Coarse loamy	1.0 to 1.5
III	Fine loam	0.75 to 1.0
IVa	Clayey (1:1 type)	0.3 to 0.75
IVb	Clayey (2:1 type)	Unsuitable

Source: Kleiss and Hoover, 1986

Nutrient Content

Recycled water contains both dissolved and suspended materials. This loading of materials can have a significant impact over an extended time due to the total volume of water which is applied.

Recycled Water Advantages

Recycled water may have appreciable nutrient content.

The water may have an appreciable nutrient content. Forms of nitrogen (N) present may include NH_4^+-N, NO_3^--N, and organic nitrogen. Other macronutrients present may include phosphorus (P), potassium (K), calcium (Ca), magnesium (Mg), and sulfur (S). These vary seasonally, and concentration will depend on the source of the water and the type of treatment facility. Additionally, several of the micronutrients also may be found in recycled water; boron (B), chlorine (Cl) , copper (Cu) and zinc (Zn) are often present. All of these may provide some turf fertilization.

Recycled Water Disadvantages

Excess nutrients may be applied to cool–season grasses at the peak time of water use, under hot and dry weather conditions when they are not needed, or to warm–season grasses at fall overseeding transition.

One of the biggest problems is that of timing of nutrient applications. Excess nutrients may be applied to cool–season grasses at the peak time of water use, under hot and dry weather conditions when they are not needed, or to warm–season grasses at fall overseeding transition. In each case stress tolerance and pest management effectiveness can be adversely affected.

Effluent may provide significant amounts of nitrogen, especially if large volumes of recycled water are being applied. However, this may be seasonally dependent. A major concern with the application of nitrogen is that the site assimilative capacity may change over time. With golf course turf, nitrogen, which is not volatilized, is not actually removed from the plant–soil system except on areas where clippings are removed. Permanent storage will occur in the soil as organic matter accumulates, but every soil has a dynamic point at which soil organic matter accumulation changes. Once this equilibrium of nitrogen loading, fixation and mineralization occurs, the soil's assimilative capacity changes. The time required to reach this point varies depending on the soil texture, type of turf and environmental influences. Close monitoring is required to determine if nitrogen is moving off–site or downward into ground water.

Cost

Cost is a factor with both direct and indirect components. The direct component is obvious when measured as costs per acre–ft, cubic feet, or thousands of gallons. The indirect costs are less obvious and must be determined on the basis of how the use of recycled water forces changes in management programs.

The indirect costs are less obvious and must be determined on the basis of how the use of recycled water forces changes in management programs.

Recycled Water Advantages

Direct cost of recycled water versus potable water is readily obvious. The cost differential can be up to ten times less expensive for effluent, thereby providing a substantial savings. The cost of pumping well water to the irrigation source often is up to four or five times the cost of recycled water. After the water from either source reaches the facility, other factors involved in irrigation are similar.

Recycled Water Disadvantages

Many of the disadvantages pertain to the indirect costs. If there is an abundance of high quality surface water from a lake or pond immediately available as the irrigation source for example, then the cost of switching to an effluent source is a direct one. However, there may be a way to factor some of the salary expenses involved with irrigation management into the effluent disposal, thereby allowing a savings in golf course maintenance costs.

Disadvantages associated with costs:
- *restrictions on where and when wastewater is applied*
- *abundance of high quality surface or groundwater nearby*
- *monitoring of potable and wastewater for problems*

Most of the costs associated with the application of effluent are in the category of less obvious indirect costs.

These may include restrictions on where, when, and how often effluent is applied, and its method of application. Restrictions on "where" may necessitate more than one delivery system. If local regulations restrict application areas based on use, then two independent systems may have to be installed to allow selective application within a given location. Similarly, if application time is restricted, there may be an extra labor requirement involved. Irrigation management is a very time–consuming, intensive operation under any condition, but even more so when dealing with regulatory issues. Under certain situations, there may be a need to irrigate when turf demands are low. This is an added indirect cost for irrigation system operation and turfgrass management, as well as a very real consideration both financially and agronomically.

Regardless of the situation, monitoring of the site will be required. The degree will be dependent on local regulations, and will include surface water where storage runoff may be a factor. If ground water is involved, the installation of monitoring wells may be required. Monitoring may require that certain parameters be evaluated at certain frequencies with local regulations dictating the exact scheduling. Analysis of the water will have to be performed by a certified testing laboratory, and there is a cost involved with this sampling.

Water Quality

A number of parameters must be considered in dealing with recycled water quality[19]. While the nutrient load is a factor in water quality and has been listed as an advantage, many of the nutrients also are salts, which means they influence total soluble salts (TSS) or water salinity. The most desirable situation is to have a water effluent with as little residual dissolved or suspended material as possible. However, this is rarely the case, and all adverse effects on soil and plants from recycled water would be listed as disadvantages.

Recycled Water Disadvantages

Water quality varies greatly depending on the source and the efficiency of the treatment system. Water quality can range from good to poor, depending upon the composition. This may require additional inputs to the management program to compensate for this impact.

Water quality analysis should include the concentrations of suspended solids, specific ions and pH, total soluble salts and calculation of a leaching requirement, calculation of the Sodium Absorption Ratio (SAR), Biological and/or Chemical Oxygen Demand (BOD, COD), toxic materials (particularly Volatile Organic Compounds [VOCs]), and total or fecal coliforms (Table 2). Local regulations may also dictate additional analyses. Problems associated with each of these and their potential effects on turf management are as follows:

Water quality disadvantages:
* *suspended solids*
* *specific ions*
* *pH*
* *total soluble salts*
* *leaching requirements*
* *Sodium Absorption Ratio*
* *biological or chemical oxygen demand*
* *toxic materials*
* *total or fecal coliforms*

* Suspended solids

  These can accumulate on the surface and cause sealing of the soil. This is especially critical if the accumulation includes a substantial mineral component. If the solids are organic, they may be decomposed by soil microorganisms if soil temperatures favor microorganism activity. Accumulations of materials in surface zones during cooler weather, with subsequent decomposition during warmer weather, may lead to periods of extreme oxygen depletion in the soil if, as is the usual situation, decomposition occurs at very shallow depths where the root system may be concentrated. Rooting and turf vigor then would be adversely affected. Suspended solids are filtered and potentially decomposed at the soil surface if soil temperatures are greater than 60^0 F and if the soil is unsaturated and permits good oxygen exchange[2].

  Another consideration is the effect of suspended solids on highly modified soils. Where root zone mixes consist of a high sand content specifically to ensure good drainage and compaction resistance, an increase in the amount of both organic and mineral suspended materials would potentially plug

Suspended solids can accumulate on the surface and cause sealing of the soil.

Table 2. Summary of Recycled Water Quality Components Which May Affect Turf Management Decisions and Their Impact

Parameter Analyzed	Potential Impact	Turf Management Consideration
Suspended solids	Soil sealing	Increased coring
Total-N, NH_4^+-N, NO_3^--N, Organic-N	Build–up of nitrogen; Untimely availability	Controlled nitrogen fertility
P & K	Contaminated runoff to ponds/lakes	Control runoff, monitor soil levels
Ca, Mg, Na	SAR high; Soil dispersal	Increased calcium applications
pH, carbonates, bicarbonates	Increased soil pH; Affects nutrients	Acidification of water or nitrogen sources
BOD/COD	Organic loading depletes soil oxygen	Increased coring, improve root growth
Conductivity (TSS)	Accumulation of salts in root zone	Leaching with irrigation
B, Cl^-, SO_4^{2-}	Potential specific ion toxicity	Monitor and offset with fertilization
Heavy metals	Toxicity to plant roots	Monitor and precipitate with P
Toxic materials	Toxicity to plant	Monitor and leach treat with charcoal
Total/Fecal coliforms	Human exposure	Monitor and isolate contact

the macropores, thereby altering the soil profile and the infiltration and percolation rate. One management strategy to overcome the effects of suspended solids is to increase the number of corings (aerifications).

The presence of suspended solids can affect irrigation system operation by plugging sprinkler head openings and valves as well as being abrasive to plastic and metal components.

The presence of suspended solids can also affect irrigation system operation by plugging sprinkler head openings and valves as well as being abrasive to plastic and metal components. This will affect irrigation system component life expectancy and could be a substantial additional cost.

• Nitrogen loading

All inorganic nitrogen is immediately available for turf uptake. If not taken in by the plant or used by the microbial population, it will leach. The management program will be affected if the plants are over stimulated and become soft and succulent. With organic nitrogen, as little as 20 percent or as much as 50 percent may be mineralized in the first year. This means a significant amount of nitrogen could build up and then be released during the warmest time of the year when soil temperatures allow for unlimited microorganism activity. This makes it very difficult to control turf fertility, especially after a few years of organic nitrogen accumulation.

A significant amount of nitrogen could build up and then be released in the warmest time of the year when soil temperatures allow for unlimited microorganism activity.

• Phosphorus and potassium

These are not normally a problem unless the recycled water is being held in a retention pond for an extended length of time. If so, phosphorus and nitrogen favor algal blooms which become a significant pond/lake management problem. The amount of phosphorus and potassium added would not exceed what the soil can assimilate and the turf requires; in fact, it is rare that the effluent can add, and the soil retain, adequate potassium for the annual requirement. Similarly, the amount of phosphorus typically added will not affect the availability of other nutrients in most soils, and may add enough under certain situations for good turf growth.

Phosphorus and nitrogen favor algal blooms which become a significant pond/lake management problem.

• Calcium, magnesium, sodium

The major concern with calcium and magnesium is not the total amount of these elements, but their relationship to sodium content and the resultant Sodium Absorption Ratio (SAR). Excess sodium displaces calcium on the soil exchange sites, causing deflocculation of the soil structure, or soil dispersal. This can result in increased compaction, thereby limiting oxygen exchange and affecting rooting and turf vigor. When the SAR is greater than 10, a recommendation normally is made to apply calcium, usually as gypsum, and to apply excess irrigation to displace and leach the sodium. However, there are costs involved. For example, gypsum requirements and costs may be high, and there will be a requirement for excess irrigation to leach. All of this requires labor, electricity for pump operation, and increased wear on the irrigation system.

Excess sodium displaces calcium on the soil exchange sites causing deflocculation of the soil structure, or soil dispersal.

• pH and total carbonates

Bicarbonates and carbonates both affect the pH of the recycled water and potentially the soil chemical properties. These are measured in milliequivalents per liter and are a source of alkalinity which may affect the water and soil. Water quality analyses commonly report total carbonates (bicarbonates and carbonates) since both affect pH levels. If the amount is significant, greater than (>) 2.5 meq/l (150 ppm), the pH of the soil may be affected with long term recycled water use. This in turn affects nutrient availability. To offset this effect, one may use either acid forming nitrogen fertilizers, acid-injection into the irrigation water, or sulfur applications to the soil.

Bicarbonates and carbonates both affect the pH of the recycled water and potentially soil chemical properties.

• Biological/chemical oxygen demand

This represents the organic and microbiological load in the effluent. It is a determination of the amount of oxygen required for decomposition to occur. The most accurate analysis is the Chemical Oxygen Demand (COD) since it represents the maximum potential oxygen requirement once the organic materials are deposited in the soil. Since the microorganism population in the soil is so diverse, organic materials added in the recycled water stream become an energy source for microorganisms, and therefore possess an oxygen requirement for their metabolism. This can result in subtle but measureable reductions in growth under low BOD/COD concentrations[8]. If the organic loading creates a situation where the microorganisms utilize oxygen at a greater rate than the exchange capacity of the soil, it may result in oxygen depletion and an interruption in root function[3].

If the organic loading creates a situation where the microorganisms utilize oxygen at a greater rate than the exchange capacity of the soil, it may result in oxygen depletion and an interruption in root function.

- Conductivity (total soluble salts)

Salinity problems primarily manifest themselves in three ways: (1) osmotic effects, (2) accumulations of specific ions, and (3) their effect on soil physical conditions[14]. While most turf species used on golf courses are reasonably salt tolerant, irrigation of golf course areas usually includes plants other than grasses. Ground covers, ornamental bedding plants, shrubs, and trees are examples. The relative salinity tolerance of these grasses and plants may cause a shift in the components as one species is favored by saline irrigation over another. Leaching of salts from the root zone is critical to maintaining turf under saline irrigation. The conductivity of soils can be *two* to *ten* times higher than that of the irrigation water[7]. This demands constant attention to ensure that adequate leaching occurs. On sandy soils, leaching of salts is easily accomplished with excess irrigation. On heavier textured soils, this becomes more difficult in that larger volumes and longer irrigation times are required. Also, since many of these salts may become attached to the collodial complex of the soil, their leaching potential is lower and more intensive irrigation is required.

Total soluble salts
- *osmotic effects*
- *accumulation of specific ions*
- *effect on soil physical properties*

- Specific ions

Boron, chlorine, and sulfates all may become toxic when absorbed by the plant if concentrations in the irrigation water are excessive. Recent infor-

Boron, chlorine, and sulfates all may become toxic when absorbed by the plant if concentrations in the irrigation water are excessive.

mation suggests that since boron accumulates in the leaf tips in situations where clippings are removed, high boron levels do not create severe problems. However, where clippings are returned there is concern about leaf burn and boron buildup in the soil. Boron concentrations are excessive when they exceed 1 to 2 ppm. Boron is difficult to leach, requiring twice as much water to leach as other soluble salts. Boron also is of critical concern to other landscape plantings such as woody ornamentals, which are not as boron–tolerant as turfgrasses. Some plants are sensitive to boron levels as low as 0.33 ppm.

Where good drainage is available and the leaching potential is high there is less concern with chlorine and sulfates because both of these salts have good water solubility. However, for some salt sensitive grasses and other plants, levels of 250 to 400 ppm is considered undesirable for irrigation[1].

* Heavy metals

Heavy metal levels should be monitored periodically, both in the water and soil.

These include copper, nickel, zinc, lead, chromium, mercury and arsenic. While exact levels at which these become a problem, either separately or collectively, are unknown, the levels should be monitored periodically both in the water and soil. Many of these metals complex with phosphorus and other elements to make them biologically unavailable. As soil levels build, it may be necessary to increase phosphorus applications in order to complex these ions and keep them unavailable.

* Toxic organics

Many of these are also known as Volatile Organic Compounds (VOCs) such as toluene and xylene. They may create direct toxicity problems if present in high concentrations.

* Fecal coliforms

These are a health concern related to human exposure and are not an agronomic consideration.

All of these water quality factors potentially affect the turf cultural program. Water quality affects turf fertility, as more potassium may be needed to offset the effects of sodium, and it may be more difficult to control nitrogen levels. It is easy to overirrigate, especially if leaching is required, and excess suspended solids, organics, bicarbonates, and sodium may then present soil problems.

Health Considerations

Requirements for tertiary or advanced treatment before use for irrigation varies from state–to–state. Unless advanced treatment is required, there exists the possibility that some agent which could affect human health may be filtered by the turf or soil surface rather than eliminated at the treatment plant.

If unwanted exposure to pathogens or toxic substances were to occur, then the possibility of health effects including infection, disease, hypersensitivity, acute toxicity, and cardiovascular–immunological reactions may occur[4,12]. In most situations this is greatly reduced because the local health regulations require the water to undergo tertiary treatment[18]. However, one of the biggest disadvantages is not necessarily a physical problem, but a largely psychological one. The public simply objects to the use of this resource where public exposure can occur.

Miscellaneous problems may include
- *corrosion or plugging of irrigation system*
- *accumulation of suspended solids in storage ponds*
- *chlorine toxicity to plants*
- *interaction of the various disadvantages*

Miscellaneous Problems

These may include abnormal pH, which leads to corrosion of pipelines and irrigation system components, plugging of the irrigation system, accumulation of suspended solids in retention ponds, and similar difficulties. Additionally, where residual chlorine is required for advanced treatment, some leaf burn may occur, especially on bedding plants and ornamentals. One of the most subtle problems is the multiplicative effect of more than one disadvantage. If one problem reduces the management efficiency to 80%, then two problems reduce the efficiency as their

Despite increased problems, concerns and cost, these impacts are not necessarily insurmountable, rather they present a challenge which demands attention to every agronomic and management detail.

product; i.e., 80% x 80% = 64%. So compound problems make increasing demands on the superintendent's management skills.

Summary

Use of recycled water for irrigation of golf courses presents a unique set of advantages and disadvantages. They potentially affect every decision the superintendent must make. However, despite increased problems, concerns and cost, these impacts are not necessarily insurmountable; rather, they present a challenge which demands attention to every agronomic and management detail.

References

1. Beard, James B. 1982. Turf management for golf courses. Burgess Publishing Co., Minneapolis, MN.

2. Carlile, B.L. and J.A. Phillips. 1976. Evaluation of soil systems for land disposal of industrial and municipal effluents. Water Resources Research Institute of the University of North Carolina, Report 76-118.

3. Clark, F.E. and W.D. Kemper. 1967. Microbial activity in relation to soil water and soil aeration. *In* Irrigation of Agricultural Lands. American Society of Agronomy, Madison, WI.

4. Crook, James. 1985. Health and regulatory considerations. p. 10:1-49.*In* Pettygrove and Asano (ed.) Irrigation with Reclaimed Municipal Wastewater—A Guidance Manual. Lewis Publishers, Inc., Chelsea, MI.

5. Crites, R.W. 1985. Site characteristics. p. 4:1-19. *In* Pettygrove and Asano (ed.) Irrigation with Reclaimed Municipal Wastewater—A Guidance Manual. Lewis Publishers, Inc., Chelsea, MI.

6.   George, M.R., G.S. Pettygrove, and W.B. Davis.
      1985. Crop selection and management. p. 6:1-
      18. *In* Pettygrove and Asano (ed.) Irrigation
      with Reclaimed Municipal Wastewater_A Guid-
      ance Manual. Lewis Publishers, Inc., Chelsea,
      MI. 1992.

7.   Harivandi, M. Ali, Jack D. Butler, and Lin Wu.
      Salinity and turfgrass culture. p. 207-230. *In*
      Waddington, Carrow and Shearman (ed.)
      Turfgrass. ASA, CSSA, SSSA, Madison, WI.

8.   Hunt, Patrick and T.C. Peele. 1968. Organic
      matter removal from peach waste. Agron. J.
      60:321-323.

9.   King, Larry D. 1986. Design and management
      considerations in applying minimally treated
      wastewater to land. p. 15-26. *In* Utilization,
      Treatment, and Disposal of Waste on Land. Soil
      Science Society of America, Madison, WI.

10.   Kleiss, H.J. and M.T. Hoover. 1986. Soil and
      site criteria for on-site systems. p. 111-128. *In*
      Utilization, Treatment, and Disposal of Waste
      on Land. Soil Science Society of America,
      Madison, WI.

11.   Kneebone, W.R., D.M. Kopec and C.F. Mancino.
      1992. Water requirements and irrigation. p.
      441-472. *In* Waddington, Carrow and Shearman
      (ed.) Turfgrass. ASA, CSSA, SSSA, Madison,
      WI.

12.   Kowal, N.E. 1986. Health considerations in
      applying minimum treated wastewater to land.
      p. 27-54. *In* Utilization, Treatment, and Dis-
      posal of Waste on Land. Soil Science Society
      of America, Madison, WI.

13.   Moore, Charles V., Kent D. Olson and Miguel
      A. Marino. 1985. On-farm economics of re-
      claimed wastewater irrigation. p. 9:1-30. *In*
      Pettygrove and Asano (ed.) Irrigation with Re-
      claimed Municipal Wastewater—A Guidance
      Manual. Lewis Publishers, Inc., Chelsea, MI.

14.   Oster, J.D. and J.D. Rhoades. 1985. Water man-
      agement for salinity and sodicity control. p. 7:1-

20. *In* Pettygrove and Asano (ed.)Irrigation with Reclaimed Municipal Wastewater—A Guidance Manual. Lewis Publishers, Inc., Chelsea, MI.

15. Page, A.L., Cecil Lue-Hing and A.C. Chang. 1986. Utilization, treatment and disposal of waste on land. p. 1-12. *In* Utilization, Treatment, and Disposal of Waste on Land. Soil Science Society of America, Madison, WI.

16. Smith, R.G., J.L. Meyer, G.L. Dickey, and B.R. Hanson. 1985. Irrigation system design. p. 8:1-61. *In* Pettygrove and Asano (ed.)Irrigation with Reclaimed Municipal Wastewater–A Guidance Manual. Lewis Publishers, Inc., Chelsea, MI.

17. The New Merriam-Webster Dictionary. 1989. Merriam-Webster, Inc., Springfield, MA.

18. Tompkins, Alan. 1992. Wastewater reclamation systems: choosing the right system. Golf Development Magazine. Sept. p. 4-5.

19. Westcot, Dennis W. and Robert S. Ayers. 1985. Irrigation water quality criteria. p. 3:1-35. *In* Pettygrove and Asano (ed.) Irrigation with Reclaimed Municipal Wastewater—A Guidance Manual. Lewis Publishers Inc., Chelsea, MI.

Retrofitting a Golf Course for Recycled Water: An Engineer's Perspective

Charles Z. Steinbergs, P.E.
Senior Engineer, Orange County Water District
Fountain Valley, California

Introduction

In California, interest in recent years in the use of recycled water at golf courses has risen dramatically. Approximately 1,000 golf courses are currently located throughout the state, each responsible for maintaining an average of 125 acres of green turf. When considering the state's recent drought situation and limited water supply, both the golf course industry and water resources planners are looking seriously to recycled water as a reliable source for meeting current and future irrigation needs. While information and experience is available concerning the design and construction of recycled water treatment plants and conveyance facilities, such is not the case when it comes to retrofitting (i.e., converting) an existing golf course facility from potable water to recycled water. Therefore, the purpose of this paper is to provide information based upon experience concerning the practical aspects of retrofitting golf courses for the use of recycled water.

There is not a great deal of information and experience available when it comes to retrofitting an existing golf course facility from potable water to recycled water.

The discussion presented herein is based upon experience gained from the Orange County Water District's (OCWD) Green Acres Project. At the heart of this project is a 7.5 million gallons a day (mgd) treatment plant that includes coagulation, filtration, disinfection and 2 million gallons of storage. Treated water from the storage reservoir is pumped through 25 miles of pipeline which traverses portions of Costa Mesa, Foundation Valley, Santa Ana and a small portion of Huntington Beach. These cites are well established, built-out communities with limited open space for new development. Therefore, when OCWD set out to implement the Green Acres

*The Orange County
Water District took on
the challenging task of
overseeing a recycled
water project in which
nearly 90% of the
potential users are
existing facilities
requiring retrofitting,
six of which are golf
courses.*

Project it took on the challenging task of overseeing a recycled water project in which nearly 90% of the potential users are existing facilities requiring retrofitting, six of which are golf courses.

Terms

There are a few terms with which one should be familiar when considering recycled water. Three synonymous terms that are used interchangeably within the recycled water industry are as follows: recycled water = reclaimed water = reclaimed wastewater.

*Recycled water =
reclaimed water =
reclaimed wastewater.*

These terms refer to water, normally of wastewater origin, that has undergone significant treatment to be made suitable for reuse within certain limitations. On occasion, the term "effluent" has been used as a labeling term, but it is not considered adequate because it only denotes water discharged from a treatment plant and is void of conveying the thought of the water being reused.

*Potable water =
domestic water =
drinking water.*

Another set of three synonymous terms with which one should also be familiar are as follow: potable water = domestic water = drinking water. These terms refer to water that is suitable for human consumption.

*For a golf course,
"retrofitting" could
best be depicted as the
process in which the
potable water system
is protected while
converting the irriga-
tion system for
recycled water use.*

One additional term deserving attention is "retrofit." This term can be used either as a noun or verb. As a noun it would refer to an existing facility that had undergone conversion in a structural sense from one mode of operation to a new mode of operation. The verb form of this term may be defined as action taken to reconstruct or modify an existing facility for the purpose of making it suitable for a new mode of operation. For a golf course, retrofitting could best be depicted as the process in which the potable water system is protected while converting the irrigation system for recycled water use.

Laws

There are three major state laws in California concerning recycled water that are pertinent to this discussion. The first law is referred to as AB 174 which became effective January 1, 1992. In brief, this law prohibits the use of potable water for various nonpotable uses, including golf course irrigation, if recycled water is available at a cost which is comparable to the cost of supplying potable water. The other two laws, known as Title 22 and Title 17, contain health department requirements. Title 22 sets forth standards on recycled water treatment, quality and allowable uses. Title 17 establishes requirements for backflow protection of the public potable water system.

The major state laws which concern recycled water use in California include:
- *AB174—prohibits the use of potable water for non–potable use*
- *Title 22—sets standards on recycled water treatment*
- *Title 17—requires backflow protection of public potable water systems*

Planning

A prudent step for staff at any golf course is to check with the local potable water and wastewater agencies on master planning efforts that call for the use of recycled water. The earlier the coordination efforts can start, the better for all parties concerned. For water planners, it is well known that golf courses are prime targets when recycled water projects are being considered. The six potential golf course users within the Green Acres Project service area were all contacted early during the planning phase. Experience at OCWD has shown that the "cradle to grave" effort (i.e., planning, permits, negotiations, design, construction and final switch over) in retrofitting a typical golf course can take six to twelve months to complete.

A prudent step for staff at any golf course is to check with the local potable water and wastewater agencies on master planning efforts that call for the use of recycled water.

Consultant

Once a golf course site is judged viable for recycled water use, the services of a retrofit design consultant should be considered. While it is possible to design a retrofit project in-house, experience has shown that in most cases the health department and engineering issues are complex

The health department and engineering issues are complex and numerous enough to warrant outside expertise.

Design efforts are
focused on:
  • *retrofit of the*
    irrigation system
  • *protective measures*
    to prevent cross
    connections
    between potable
    and recycled water

and numerous enough to warrant outside expertise. When hiring a consultant, their experience in doing other retrofit projects should be carefully considered. Also it is important for the consultant to have a working knowledge of health department and local water department requirements. The consultant's design efforts are mainly focused on the following two areas: (1) retrofit of the irrigation system; and (2) protective measures needed to prevent cross connections between the potable water and irrigation water systems. In most cases, the consultant needs a set of plans showing utilities. Up–to–date drawings showing the location of irrigation and potable water lines are invaluable in carrying out the design work.

Health Department Inspection

Before design work
commences, an
inspection of the golf
course is scheduled
with the appropriate
health and regulatory
agencies.

Preferably, after the retrofit design consultant is hired, but before design work commences, an inspection of the golf course is scheduled with the appropriate health and regulatory agencies. This inspection is required before any retrofitting work can take place. In California, the agencies involved include the county or city health departments, the California Department of Health Service (CDHS) and the Regional Water Quality Control Boards (RWQCB). In some cases, the CDHS and RWQCB may choose to rely on the county or city health department staff to perform the actual inspection.

The design consultant should be present for this important inspection. In most cases, the health agency issues general and site specific requirements to be used by the design consultant in preparing the retrofit plans. Plans must receive final approval by the appropriate health departments, potable water agency and recycled water agency.

Based upon experience gained by OCWD through these inspections with the Green Acres Project, the following problem areas are typical of those to be expected.

Cross Connection Protection

Some key requirements are as follows: (1) a minimum of a reduced pressure principle backflow (RP) device at the service connection to the golf course's potable water system (additional onsite RP devices will be required if the potable water system is used for incidental irrigation, (e.g., around club house); (2) a double check valve (DCV) assembly as a minimum for any dedicated fire line using potable water; (3) an air-gap assembly on potable water lines connected to lake or pond fill lines; and (4) removal of all physical connections between the irrigation system and potable water system.

Expected problem areas during retrofit projects:
- *cross connection protection*
- *common trenches*
- *mystery lines*
- *wells*
- *drinking fountains*
- *quick couplers*

Common Trenches

Recycled water lines and potable water lines are not allowed to exist in the same trench. For parallel lines, the minimum separation is 10 feet. Under special circumstances the minimum distance can be reduced to 4 feet.

Mystery Lines

Lines whose origins and/or destinations are unknown must be investigated. Lines that are shown on existing drawings as being "future" or "abandoned" may have to be field verified if such lines would pose as potential cross connection problems. Interconnections of lines between adjoining properties may have to be terminated.

Wells

Onsite irrigation and potable water wells will need to be protected if recycled water is to be used. For irrigation wells, a 10–foot concrete slab extending 10 feet in each direction may be required. Potable water wells will be subject to this requirement and numerous other conditions. If the well is not essential to the golf course's operation, then in some cases abandonment may be the most practical solution.

Drinking Fountains

Drinking fountains located on the course must be protected from recycled water spray.

Drinking fountains located on the course must be protected from recycled water spray by special covers, or they can be relocated to non-spray areas. Because standard fountain head covers are not readily available on the commercial market, innovation will be required on the part of golf course staff and the design consultant if the fountain cover approach is to be implemented. The OCWD and one consultant have jointly developed a stainless steel cover for use in the Green Acres Project which the local health authorities have accepted.

Quick Couplers

Although normally not a big problem for golf courses, use of recycled water quick couplers must be restricted to maintenance personnel. The health department will impose additional requirements if the public has access to their use. If potable water quick couplers are also used onsite, they must be of a different type from the recycled water quick couplers.

Design Consideration

A few items that are noteworthy concerning the retrofit design are as follows:

Tagging, Labeling & Painting

Design considerations
* *tagging, labeling and painting*
* *piping*
* *fire line backflow protection*
* *pond storage*
* *warning signs*

Recycled and potable water control valves and boxes will have to be identified by some means. For the valve itself, normally the method of choice is to attach a polyurethane warning tag using a nylon cable tie to the valve's control wiring. These warning tags are commercially available. Concerning the lids, if the site has relatively few valve boxes, then one option is to replace them with special purple lids and boxes. At OCWD, a hot iron with exchangeable letters has been used quite effectively in branding existing plastic box lids with either "Reclaimed Water" or "Potable Water". Painting or stenciling plastic lids is not recommended due to poor weather resistance, but may be quite satisfactory as means for carrying out identification on concrete and metal surfaces. Recycled water irrigation controller units will also need to be identified by having warning labels affixed to the housing. All above-ground or exposed recycled water lines and facilities will need to be painted purple (the color– code standard for purple is Pantone 512).

Piping

Normally, for a retrofit project only the replaced portions of buried pipe will be required to be constructed of purple plastic pipe or a non-purple colored pipe with identification tape. A serious misconception held by many who are unfamiliar with health department retrofit requirements is that a retrofit site will automatically be required to replace all of its existing irrigation pipe with new purple pipe. To date, this has not been the experience of OCWD in dealing with the health authorities on retrofit projects.

Normally, for a retrofit project only the replaced portions of buried pipe will be required to be constructed of purple plastic pipe or a non-purple colored pipe with identification tape.

Fire Line Backflow Protection

An appropriate backflow device (DCV device would be the minimum protection allowed) is required on any fire line entering the golf course. The local fire department should be contacted before a backflow device is installed on any fire line. The fire department determines if hydraulic calculations or flow test are required. This matter should be examined early in the design phase.

Pond Storage

A significant task requiring study and consideration is the storage of recycled water in open, uncovered ponds. Algae blooms can be a serious problem due to the availability of a nutrient source in the recycled water and sunlight. There are no easy solutions to this problem.

Algae blooms can be a serious problem due to the availability of a nutrient source in the recycled water and sunlight.

Warning Signs

Warning signs are required.

Warning signs are required. They need to have a purple background with white or black lettering. These signs can be used to give the golfers a positive message concerning the golf course's effort to practice water conservation. For the Green Acres Project the warning signs say, "For Water Conservation, this Golf Course is Irrigated with Reclaimed Water - Do Not Drink!" One is allowed considerable liberty in deciding on the sign's wording, provided that the phase "Do Not Drink" is included.

Health authorities usually require the recycled water warning verbiage be printed on the individual golf score cards.

As a further consideration, experience has shown that, on occasion, golfers become uncertain over whether drinking fountains and restrooms are plumbed separately from the recycled water system. To head off any consternation, it may be prudent to post a second sign (blue background with white lettering) which says, "Potable Water is Provided to Drinking Fountains and Restrooms."

Finally, health authorities usually require the recycled water warning verbiage be printed on the individual golf score cards.

Conclusion

Every retrofit project will pose at least one set of circumstances that will draw deeply on the resourcefulness of the retrofit design consultant and golf course staff.

In OCWD's experience with the Green Acres Project, every retrofit project will pose at least one set of circumstances that will draw deeply on the resourcefulness of the retrofit design consultant and golf course staff in arriving at a workable solution. One should not be surprised or anguished when the unexpected manifests itself. Retrofit projects involving recycled water are most definitely a challenge, but a challenge that can be successfully met with careful planning, good engineering and an earnest desire on everyone's part to "make it work." When considering how water supplies are becoming increasingly scarce, a golf course retrofit done today will always be looked upon retrospectively as a wise move that was well worth the effort.

Princeton Meadows Country Club

Paul Powondra
Golf Course Superintendent
Princeton Meadows Country Club
Plainsboro, New Jersey

Princeton Meadows Country Club is located in central New Jersey, midway between New York and Philadelphia, near Princeton University. The golf course is part of a planned community begun in the early 1970s and built on 1,000 acres of former farmland. The community is complete with apartments, condominiums, townhouses, single-family homes, a shopping center, an office park, a research park, and a wastewater treatment facility. The waste treatment plant (WTP) was built not only for the project, but was also tied into the existing town of Plainsboro.

The golf course is part of a planned community begun in the early 1970s and built on 1,000 acres of former farmland.

The developer was required to provide a certain percentage of open space within the project and to dispose of secondary effluent in the summer months when the effluent could not be 100 percent diverted to a nearby stream. The existence of the golf course was essential to the project, taking care of the open space and effluent water issues and providing an attractive amenity for real estate sales. The first nine holes were in place by 1976 and the second nine was opened in 1980. The eighteen holes comprise 126 acres, with 100 acres irrigated.

The existence of the golf course was essential to the project, taking care of the open space and effluent water issues.

The developer had the golf course closely tied to the waste treatment plant. The irrigation system was actually considered a WTP asset. One owner even referred to it as the fourth stage of the sewage treatment facility. The WTP covered all the capital expenditures and the costs for irrigation system maintenance, the electrical pumping costs for irrigating, expenditures for the groundwater monitoring program, lab fees, reporting requirements to the state DEPE, and 25 percent of the golf course superintendent's salary. The company accountants considered this a cost benefit to the developer since the

The irrigation system was actually considered a waste treatment plant asset.

expenses charged to the WTP were included in the sewer service rate base and, of course, could be charged back to the rate payers.

In the early 1980s the plant was upgraded to a tertiary treatment facility and expanded from 500,000 gallons per day to 1.5 million gallons per day. In 1988, the plant was required to change its name to Princeton Meadows Utility Company (PMUC), solely because one of the lines ran under a public road and the State Board of Public Utilities had to step in to control that particular issue. This change occurred even though the plant remains privately owned.

By 1987, with the Clean Water Act on the books, we started to see more required record-keeping, daily flow records had to be kept by the golf course, and minimum flow requirements were imposed upon the golf course.

My tenure at the golf course began in 1982. From 1982 to 1986, use of the wastewater was relatively easy going. We used it as we needed it, very little direction was given from the PMUC operator, and we had no minimum requirements or daily flow. By 1987, with the Clean Water Act on the books, we started to see more required record-keeping, daily flow records had to be kept by the golf course, and minimum flow requirements were imposed upon the golf course. Ultimately, by 1989 we were charged with disposing 800,000 gallons per day. In the fall of 1989, the plant operator warned me of the potential requirement to dispose a million gallons a day in 1990. About that time I was advised to clear out of there because of the potential for turf failure with so much water!

In 1990 and 1991, the PMUC and the State Department of Environmental Protection began a stream study to evaluate the effect that increased flows would have on the stream during the summer months and to investigate the impact of diverting tertiary effluent to the stream. These were happy days for me because my daily requirements were dropped to "only" half a million gallons a day. That was a number I could deal with!

In 1992, the golf course was sold. The golf course and the Princeton Meadows Utility Company went their separate ways. The rules changed. We received our own operating permit from the state for effluent use with PMUC as our supplier, and the irrigation system became ours. We had no minimum requirements to spray irrigate.

I had to learn how to manage water again by letting things dry up and firm up. It was wonderful!

The conclusion of the stream study was that the PMUC effluent was of such a quality that 100 percent of the water could be diverted to the stream year around. The utility company now only charges us for the electricity to pump and deliver the effluent through the main distribution line to our holding pond.

The utility company now only charges us for the electricity to pump and deliver the effluent through the main distribution line to our holding pond.

The sale of the golf course put the burden on us to handle the permit costs and responsibilities, the groundwater monitoring, the maintenance expenses for the irrigation system, and electricity. Now, 100 percent of my salary is a golf course budget item. We also were required to install three new groundwater monitoring wells at a cost of $8,500.

The sale of the golf course put the burden on us to handle the permit costs and responsibilities, the groundwater monitoring, the maintenance expenses for the irrigation system, and electricity.

The wastewater plant operators are to be commended for the job they do. This is an unglamorous job that one usually doesn't brag about, but it is very important. Superintendents can relate to this behind-the-scenes responsibility, although the superintendent's job is not quite that way anymore. The wastewater plant is rarely noticed by the community unless there's an odor problem. Wastewater plants are critical to community health. The operations are very technically involved and closely regulated. As a morale booster, the PMUC adopted the motto, "We're #1 in the #2 business."

Permit Requirements

The first permit issue we were required to deal with was the installation of an anemometer at the pump house. It had to be the same height as the height of the maximum sprayed wastewater and electronically connected to the pumps. To reduce the aerosols, we arranged for the pumps to shut down at wind velocities of 15 miles per hour. This cost $1,300, plus the electrician's time. It took about a day to get the anemometer hooked up, at a cost of about $40 an hour. We were not required in our permit to have a rain sensor, but we do have one

The first permit issue we were required to deal with was the installation of an anemometer at the pump house.

installed as the result of an old permit. Another golf course being constructed in New Jersey with a similar permit as ours is required to have a rain sensor.

Monitoring Data

The pumphouse discharge monitoring data are reported to the EPA.

The pumphouse discharge monitoring data are reported to the EPA. A New Jersey certified lab does the sampling and testing. A flow measuring device, installed at the pumphouse, obviated the need for other devices at the lagoons. Sampling and testing costs $95 a month, and the data is being gathered and studied to establish limits for coliform bacteria.

The New Jersey Pollution Discharge Elimination System (NJPDES) permit is intended to determine: direction of groundwater flow; hydraulic performance of disposal areas; and depth to groundwater. It also requires that we monitor groundwater quality. Several types of monitoring wells are on the property. Among these are nine water table monitoring wells. At one time these wells were used for upgradient and downgradient monitoring purposes, but today they are used strictly for monitoring the water table as required by the state.

The New Jersey Pollution Discharge Elimination System permit determines:
- *direction of groundwater flow*
- *hydraulic performance of disposal areas*
- *depth to groundwater*
- *groundwater quality*

We also have the recently installed new upgradient and downgradient wells. These two types add another nine wells to the property for a total of 18 wells. The upgradient and downgradient wells also contribute to the water table data. All the wells are sampled quarterly. Also, a contour map is constructed quarterly. An engineering firm is hired to construct the map at a cost of $250 per quarter. The permit states that the map has to be on a 8 1/2" x 11" piece of paper, so the entire golf course property must fit on this small piece of paper.

The standard values for coliform and pH are under study (Table 1). The maximum nitrate standards are ten parts per million. It is an advantage to have the regulator who supervises your permit know the history of your property. The golf course was built on land that was previously planted to potatoes and apple trees. Last summer one of the wells spiked to 12 parts per million. We thought we were in trouble, but the regulator knew the past history of the site. Spikes usually occurred in the summer and the observed spike of 12 ppm was not an issue. The monitoring and sampling costs $880 a quarter.

Table 1. Quarterly groundwater monitoring and sampling standards for the upgradient and down gradient wells

Parameter	Standards	Sampling Type
Ammonia-N	0.5 ppm	grab[1]
Coliform bacteria	under study	grab
Nitrate N	10 ppm	grab
pH	under study	grab

[1] An individual sample of at least 100 ml collected over a period not exceeding 15 minutes.

Required Signs

In our permit we have requirements for two readily visible and legible caution signs at each lagoon, plus one must be in the clubhouse. The caution sign must also be in compliance with the requirements listed in Table 2. I called the DEP trying to locate the "Right-to-Know Manual" and found that it does not exist. The symbol on the sign, referred to as Mr. Yuck, originated in California according to the person who wrote the permit.

We have requirements for two readily visible and legible caution signs at each lagoon, plus one must be in the clubhouse.

Table 2. Specific requirements for caution signs posted at lagoons and in the clubhouse

---

1. A picture of "Mr. Yuck," as found in the NJDEPE's Right-To-Know Manual. Mr. Yuck is intended to discourage people from drinking effluent or swimming in the lagoons.

2. A written warning that the lagoon and irrigation water is unfit for human consumption.

3. A written description of the lagoon and irrigation water as treated effluent and rainwater.

4. A written notice of the NJDPES permit which allows for wastewater discharge on the golf course. Current copies of both the Princeton Meadows Country Club's NJPDES permit and the PMUC's NJPDES permit must be kept at the clubhouse at all times and made available for public inspection.

The symbol on the sign, referred to as Mr. Yuck, originated in California according to the person who wrote the permit.

No other specific requirements were made for the sign, so I came up with my own. My prototype sign was 18 inches x 24 inches, with Mr. Yuck's diameter at 6 inches, using white letters on a blue background (Table 3). The state has asked us to modify the wording to include that the permit is available for review at the clubhouse.

The spray irrigation system is required by state statute to be under the supervision of a licensed wastewater treatment plant operator who meets the New Jersey requirements.

The spray irrigation system is required by state statute to be under the supervision of a licensed wastewater treatment plant operator who meets the New Jersey requirements as the "person in charge." Even the person who wrote our permit feels this is unnecessary, but we will have to live with it for now. We have hired a licensed operator for $200 a month which has worked out to be a bargain. Normally, the work requires approximately two hours a week, primarily for the paperwork. The state is comfortable that the person in charge is technically responsible.

The spray irrigation water and the ponds on this golf course are composed of treated effluent and rainwater.

This water is unfit for human consumption.

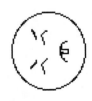

Princeton Meadows Country Club operates under a permit issued by the NJ Department of Environmental Protection and Energy which allows for the application of wastewater on the golf course.

Table 3. A prototype of the Princeton Meadows Country Club caution sign

Record Keeping

Record keeping requirements:
- *inspect wells for signs of damage*
- *inspect irrigated areas for ponding, odors, or an overabundant loss of vegetation*
- *daily wind speeds, spray activity, and gallons pumped*

As with any permit there are record keeping requirements. On a monthly basis we are required to inspect all wells for signs of damage and to assure their structural integrity. The inspector needs to maintain records in the log book indicating the date of inspection, inspector's name, and the condition observed. We have found that it is important to do more than just indicate "Okay." It is beneficial to write something that indicates you did go around and look at the wells.

The weekly inspections are conducted on the immediate area and the area surrounding the disposal areas for evidence of problems, such as ponding, wet areas, odors, and an overabundance or loss of vegetative cover. Daily records must be kept on wind speeds and spray activities in addition to recording gallons pumped per day. The weekly and daily records are kept during the months the spray irrigation system is in use. Some records, such as the definition of spray activities and just how wind speeds are to be recorded have been left vague. We record the wind speed once in the morning when we are making a record for the last 24 hour water flow.

An important point for us has been coopera- tion with the treatment facility personnel and the state regulators.

A lot is involved with this project, but we are working our way through it. An important point for us has been cooperation with the treatment facility personnel and the state regulators. Communication has been, and will remain, a priority for smooth operations.

Kiawah Island Golf Club

George Frye
Director, Golf Course Maintenance
Kiawah Island, South Carolina

Introduction

Kiawah Island Golf Club is a 10,000 acre resort located on a barrier island off the coast of South Carolina. The original master plan included 81 holes of golf, with a utility plant designed to process the effluent on–site. At the current time, we have 72 holes, with future plans to build another nine holes.

Kiawah Island Golf Club is a 10,000 acre resort located on a barrier island off the coast of South Carolina.

On Kiawah, effluent is considered wastewater, recycled water, or sewer plant effluent. The utility plant is obligated to supply water to the golf courses, but sometimes they do not have enough. We are obligated to take that water, but sometimes it is not needed. It is important to understand the goals and objectives of both operations. Our goals are quality turf, quality playing conditions and the disposal of the island's secondarily treated wastewater.

The Ocean Course was designed with a recycling system where effluent applied on the back nine is re–captured and recycled to the front nine. This is a form of treating the effluent and also a form of water conservation.

The disposal of this water begins at the utility plant where the actual used–potable water enters the plant. The water goes through a series of aeration chambers where the dissolved oxygen levels are maintained at two to five parts per million. From there the water enters a contact chamber where it is treated and placed in storage for use on the golf courses.

The water goes through a series of aeration chambers where the dissolved oxygen levels are maintained at two to five parts per million.

The variable speed pumps at the treatment plant are designed to send water through an 18–inch line anywhere from six to eight miles, depending on which golf course is receiving the water. These lines are laid along the parkway on Kiawah Island.

We utilize a wet well which is designed to hold 45 to 50 thousand gallons.

Kiawah utilizes a wet well which is designed to hold 45 to 50 thousand gallons. This is a closed–in system that could be recommended to anyone thinking about utilizing secondary effluent as a water source. A float system is also used which allows for control of the water level inside the wet well. When the level drops, a valve allows the water to enter. From this point, a telemetering system sends information back to the utility plant and signals the effluent pump station to send water to the golf course as necessary.

It is important to look at some of the design criteria and prerequisites that were established by the state regulatory commission as part of the entire system. In this case, the Department of Environmental Health and Environmental Control (DEHEC) was the state agency responsible for regulations.

It is essential to have buffer zones when utilizing effluent water.

It is essential to have buffer zones when utilizing effluent water. We have a buffer zone with actual setbacks from the property lines and also preferred and required setbacks where housing eventually will be built. The buffer zones will vary, depending on the kind of environment.

An irrigation system should be designed to deliver effluent from the outer fairway extremities to the inner portion using full circle as well as half circle heads, thus confining effluent spray to designated areas.

After the irrigation systems were designed and installed, monitoring wells were located strategically throughout the golf courses.

After the irrigation systems were designed and installed, monitoring wells were strategically located throughout the golf courses. The monitoring parameters were developed by the DEHEC and water samples are taken semi–annually and tested for orthophosphates, nitrates, pH, specific conductants, as well as water table elevation. More frequent sampling is done if problems or negative trends develop.

Once the basic structure for delivery, irrigation, drainage, and monitoring has been established for using effluent, it is important to look at affordability. Is the water that you are getting ready to use affordable? On Kiawah Island we have three water sources: effluent, potable and deep well. Our primary source of irrigation water is effluent.

The development of the golf course was designed around the use of effluent on Kiawah Island. Certainly, our second source of water is deep well, and lastly, potable water, which is called "liquid gold" on Kiawah Island.

In 1992, effluent water cost 35 cents per thousand gallons and potable water was $2.05 per thousand gallons.

Certainly the costs associated with the utilization of the three water sources must be considered. In 1992, effluent water cost 35 cents per thousand gallons and potable water was $2.05 per thousand gallons. The deep well ran about 95 cents per thousand gallons. In 1991, we used 155 million gallons of effluent water, which cost approximately $54,000. The potable water was very expensive— 59 million gallons cost $121,000 in 1991. Basically, our total usage was about 35 percent to 40 percent effluent water. The total cost that year was $294,000 for a total usage of 340 million gallons.

Our total usage was about 35 percent to 40 percent effluent water.

Annual rainfall has a very large impact on Kiawah Island. Forty-five inches of rain fell in 1991, but rainfall was nearly 70 inches in 1992! Too much rainfall becomes a problem because we must still take the effluent water. That certainly has to be a consideration when planning storage area for the effluent. In the month of January 1991, we had 9 1/2 inches of rain. In 1992, we received nearly 8 inches. During those months, we had trouble using all the effluent water while trying to maintain the playability of the course.

The problem we have with a lot of rainfall is that we must still take the effluent water.

The idea of injecting surplus effluent into an aquifer for storage, and reutilizing it during our growing season may be considered as future growth continues. As the island grows, we are going to have more trouble storing the excess effluent during the winter months. This will be true especially during years with wet winters.

On the back nine holes of the Ocean Course we use approximately 300 recycled gallons a minute. Our daily demand on the Ocean Course is about a million gallons a day under a full irrigation cycle. The irrigation systems are designed to apply about one acre inch a week. The DEHEC regulations stipulate that we can apply up to two acre inches per week, if needed.

Drainage is a prereq-uisite in utilizing a water like this.

The next subject that needs to be considered is the playing quality of the golf course. Effluent water plays a major role in the turfgrass quality that is maintained on Kiawah. At times, it is difficult to use an alternative source of water that is very poor quality. Consider the deep well water that is the only alternative to the potable and effluent water. The bicarbonates are over 1,000 parts per million, the soluble salts are over 1,500 parts per million, and the SAR is extremely high. Good drainage and good air movement through the soil profile are prerequisites if poor quality water is to be utilized.

The combination of nutrient–rich effluent and warm, deep well water promotes the growth of blue–green algae.

Another irrigation system problem that must be faced concerns mixing the deep well water, which comes out of the ground at 99 degrees, with the effluent water. The nutrient–rich effluent combined with the warm, deep well water promotes the growth of blue–green algae which can clog irrigation lines. This has not resulted in an algae problem on the tees and fairways, but at times the putting greens have required additional management to control algae.

In combination with potential algae problems, another concern is total suspended solids. We process an esti-mated seven to eight thousand pounds of suspended solids a year. In an effort to deal with suspended solids, the utility plant has recently installed a sludge processing facility. From there, the water is separated and transferred to a settling bed. After composting we hope to use this material for the nursery stock or flower beds on the island.

Course sand topdressing is one of the most important things that can combat the effects of the suspended solids that are going out on the golf course.

How do we best utilize effluent with suspended solids when mixing with the deep well water that is very high in soluble salts? Course sand topdressing is one of the most important things that can combat the effects of the suspended solids that are going out on the golf course. The suspended solids, to some degree, can plug the soil pore space. If soil pore spaces are plugged, that results in very little air movement, a lot of algae growth, and a poor quality putting surface.

Oxygen is obviously a prerequisite for good quality turf. I would tend to recommend putting green root zone mixes which approach the higher infiltration limits. The larger pore space allows better oxygen movement, better channels to move the suspended solids down through the profile, and also better microbial activity. Coarse sand is the key to allow the suspended solids to move through the profile. It can provide better channels for air movement, the leaching of salts, and enhanced microbial activity.

I would tend to recommend putting green root zone mixes which approach the higher infiltration limits.

Summary

Water is one of our most valuable resources. Water quality is crucial to our environment and it dictates our future. The successful results of using recycled water can be very satisfying.

The Use of Effluent Irrigation Water: A Case Study in the West

Mike Huck
Mission Viejo Country Club
California

History

The course began irrigating it's turf with secondary effluent in the early 1970's.

The Mission Viejo Country Club was built in the late 1960's with the use of effluent in mind. The course began irrigating it's turf with secondary effluent in the early 1970's. Since the spring of 1991, tertiary treated effluent has been used.

Problems

Turfgrass Quality/Salinity/Drainage Problems

Problems:
- *quality/salinity/ drainage*
- *regulations*
- *storage ponds and lagoons*
- *irrigation time restrictions*
- *equipment deterioration*

Salt accumulation and poor internal drainage of the existing putting greens, which contained 90 percent *Poa annua,* had posed problems in the past. An arid climate, along with the poor salt tolerance of *P. annua,* left turf quality of the putting surfaces far less than desirable by the end of many summers. Many of the old putting greens do not have a soil mix of uniform depth, and shallow areas stay wet while deeper areas are dry. Such conditions make these putting greens difficult to effectively leach the salt contained in the effluent. Problems occur with trees, shrubs and ground cover where damage is incurred by excessive salt buildup and lack of internal drainage of the native clay soil.

Regulations

Prior to accepting tertiary effluent in 1991 a complete

inspection of the property was conducted by the Orange County Health Department and Moulton Niguel Water District.

The primary focus of this inspection was to insure that there had been no cross connection between potable and reclaimed services and that proper backflow prevention of potable services were in place. Other regulations concerning the proper notification of players and workers had to be addressed. All part–circle sprinklers were also inspected on the perimeter of the golf course to assure that no reclaimed spray would leave the property.

The primary focus of this inspection was to insure that there had been no cross connection between potable and reclaimed services and that proper backflow prevention of potable services were in place.

Maintaining Storage Ponds or Lagoons

Due to the high nutrient content of the effluent, pond weeds and algae have become a problem along with the accumulation of sediment that results in the control of these weeds.

Other Problems

Restrictions were placed on the hours during which irrigation may occur. The district regulates that no irrigation may begin before 9:00 p.m. and that all irrigation must end by 6:00 a.m.

Maintenance equipment deteriorates much more rapidly due to constant exposure to the saltier water.

Maintenance equipment deteriorates much more rapidly due to constant exposure to the saltier water, a problem that is exaggerated when effluent is used to wash the equipment.

Monthly overseeding with creeping bentgrass altered the 90 percent P. annua greens to approximately 60 percent bentgrass over the past 3 1/2 years.

Solutions

Turf Quality/Salinity/Drainage Problems

In this case, alternating SR1020 and Pennlinks creeping bentgrasses in a monthly overseeding program altered the 90 percent *P. annua* greens to approximately 60 percent bentgrass over the past 3 1/2 years. Overseeding,

Overseeding, leaching, an organic nitrogen source to reduce salts and a sulfur source discouraged P. annua

leaching, a fertilizer program using an organic nitrogen source to reduce salts, and a sulfur source to discourage *P. annua* has been relatively successful in keeping turf on the greens in playable condition even during the worst of summers.

After attempts to improve internal drainage with the Verti–drain and Hydroject failed, the old practice putting green was reconstructed. The new green was built to USGA specifications and seeded with SR1020 and Pennlinks creeping bentgrass.

Regulations

Compliance with regulations is the only solution if use of reclaimed water is desired!

Compliance is the only solution if use of reclaimed water is desired! Issues that had to be addressed included proper marking of all ponds, irrigation controllers, remote control valves, quick couplers, new piping and even the scorecard, which now warns that reclaimed water is being used. Proper backflow prevention devices had to be provided at all potable sources and hose bibs. Hose bibs are not allowed on any reclaimed source.

Irrigation controllers needed to be labeled, in English and Spanish, with the District's operational guidelines which state as follows:

1. All irrigation must start no earlier than 9:00 p.m. and end no later than 6:00 a.m.

2. Ponding and runoff is prohibited.

3. Failure to comply will result in termination of reclaimed service.

Copies of irrigation blueprints were supplied to Orange County Health, Moulton Niguel Water District, and the San Diego Regional Water Quality Control Board.

When the club decided to pursue a ryegrass overseeding program the Water District required that the golf course would have to be closed to irrigate during daytime hours for the germination process.

Protection of the drinking water coolers on the golf course had to be provided. To access the spigot and paper cups, stainless steel cabinets which house two water coolers and have small doors that close by gravity, were designed.

Protection of the drinking water coolers on the golf course had to be provided.

Storage Ponds and Lagoons

The floating algae mats were partially controlled by using aerators. Control of duckweed and algae was marginally successful with the use of registered chemicals. Both species reappeared and increased rapidly within just a few days of treatment.

Other Problems

Stainless steel cabinets were specified for the new system's controllers to prevent corrosion.

The new irrigation system installed in 1990 was designed with features needed to stay within operational guidelines. The system provides ample pumping capacity along with adequate pipe sizing to handle the increased flow within the reduced time frame. Stainless steel cabinets were specified for the new system's controllers to head off the corrosion problem before it began.

Economic Consequences and Public Concern

In this case, the added expenses required to meet regulations were offset by the difference in the cost of effluent water. Currently the annual effluent water bill is approximately 35 percent less than the cost of potable water. It is felt that as the demand for reclaimed water grows, the price of reclaimed may reach or exceed that of potable water.

The added expenses required to meet regulations were offset by the difference in the cost of effluent water.

In California there is not much question about whether reclaimed water is "acceptable" for irrigation or not, the correct term is available! Assembly Bill 174 from the California State Legislature's Regular Session of 1991 mandates the use of reclaimed water if it is available at a reasonable cost and if the reclaimed water is of adequate quality that will not adversely affect downstream water rights, will not degrade water quality and is determined not to be injurious to plant life, fish, and wildlife.

Advantages & Disadvantages

Advantages

1. An uninterrupted supply of irrigation water.

2. The potential for saving money ($176 per acre foot in this case).

3. Returns filtered water back to the ground water supply.

4. Reduces the need for supplemental nitrogen fertilization.

Disadvantages

1. Increases the deterioration rate of maintenance equipment and irrigation system.

2. Increased salt levels in the soil reduce growth and affect the rate of recovery of plant materials and turf which, in turn, may lead to an increased susceptibility to heat stress, disease, and overall possible decline of the plant.

3. Uncontrolled growth due to excessive nitrogen in effluent.

4. Disease problems due to excessive nitrogen.

Summary and Recommendations

Since California requires the use of reclaimed water where it is readily available, there is no longer an argument pro or con.

If effluent water is your only irrigation source, as is the case with many in California, it might call into question the value of pursuing the construction of a new golf course. The response should be an affirmative one. You just need to be prepared to deal with the regulations affecting effluent water use. The mission then becomes how to make the best use of the resources you have to work with and to be realistic about the effects effluent may have on the overall success and expectations of the project.

Since California requires the use of reclaimed water where it is readily available, there is no longer an argument pro or con.

- Provide an irrigation system which can meet the demands of a reduced time frame for watering and be prepared to periodically leach the soil profile.

- Provide good internal and surface drainage to accommodate the leaching process. This applies especially to putting greens.

- To discourage uncontrolled weed and algae growth, avoid using ponds and lakes (ornamental or storage) if possible. When ponds and lakes must be used for irrigation storage, provide adequate aeration and "turn over" the water in a relatively short time.

- Select only turfgrasses, trees, shrubs and ground cover species which are salt tolerant and well adapted to the growing environment. Consider the climate, soil types, and water quality in each specific situation.

Keys to success:
- *a good irrigation system*
- *good internal and surface drainage*
- *avoid large ponds and storage lakes*
- *select salt tolerant species*

By making these prudent decisions when designing, constructing, and planting the golf course, an exceptional and successful project can be achieved.

Barton Creek Wastewater Case Study

Tim Long
Barton Creek Country Club
Austin, Texas

Introduction

The project lies in the environmentally sensitive Barton Creek watershed, which contains approximately 80,000 acres.

Barton Creek is a 4,500 acre master–planned community located in the Central Texas Hill Country near Austin, Texas. The project lies in the environmentally sensitive Barton Creek watershed, which contains approximately 80,000 acres.

At one time it was planned to spray irrigate all of the treated sewage water generated by the sub-division onto golf courses and parklands.

Originally designed as a self–sufficient residential and commercial development, changing water quality regulations during the past four years have changed requirements for irrigation on the site. At one time it was planned to spray irrigate all of the treated sewage water generated by the sub-division onto golf courses and parklands. That plan may be in jeopardy due to requirements to spray irrigate storm water to protect water quality in near by Barton Springs. Spray irrigation of effluent is regulated by the Texas Water Commission (TWC) while storm water runoff and non point source pollution abatement are regulated by the Lower Colorado River Association (LCRA), and by the City of Austin. A recent ordinance approved by the City Council in 1991 disallows the use of transfer credits if effluent irrigation is used on new golf courses. Basically the City of Austin is discouraging the use of effluent water for irrigation.

Approximately one quarter of the golf course site was previously used as an effluent disposal site before the golf courses were constructed.

Approximately one quarter of the golf course site was previously used as an effluent disposal site before the golf courses were constructed. It was natural to assume then, that it would be easier to manage the disposal of secondary treated effluent onto golf courses supplied with an additional foot of loam soil, computerized irrigation system and a well–managed turf.

Regulatory Requirements

Before irrigation with effluent was started, fairly extensive soil analysis was conducted to establish background base level information. The Texas Water Commission only requires annual soil testing for pH, total nitrogen, potassium, phosphorus, and conductivity.

During the approval process, one of the programs in which both the LCRA and TWC were interested was the use of computerized irrigation. They were concerned not just about disposal rates, but slope and soil conditions and how these affect surface runoff. They also were interested in how the irrigation system could be used to prevent over watering and subsequent leachate contamination of groundwater. One advantage in using the Toro Network 8000 system is the ability to identify site factors such as soil type, turf type and slope below individual heads when developing an irrigation managment plan. Slopes greater than 20 percent are not considered desirable for spray irrigation due to potential runoff or infiltration problems. With the Toro system, this information is programmed into the central computer which helps identify critical areas and break up watering times to prevent runoff. The LCRA and TWC also were excited about seeing weather data collected on site and used to calculate actual turf watering requirements or evapotranspiration. Additionally, rain gauges were installed to shut down the system automatically if it began to rain.

Other regulations required by the TWC are: (1) irrigation should not continue on saturated soils; (2) irrigation water should not be allowed into golf course waterways. There also must be warning signs in both English and Spanish stating: "Do Not Drink The Water" and "Stay off pond slopes where the liner may become damaged." The TWC makes regular inspections to verify the system is operating properly, usually checking all irrigation heads near streams or critical environmental features. They also may require pressure testing of transfer lines to certify that effluent isn't leaking directly into the ground water.

Before irrigation with effluent was started, fairly extensive soil analysis was conducted to establish background base level information.

Slopes greater than 20 percent are not considered desirable for spray irrigation due to potential runoff or infiltration problems.

Other regulations:
- *no irrigation on saturated soils*
- *no irrigation water allowed in waterways*
- *warning signs*

249

Disposal Rates

Application rates increase from east to west across the state. Austin, in Travis County, is allowed a maximum application rate of 2.7 acre feet per year (Figure 1). The 2.7 acre feet amount is the net effluent irrigation needed after rainfall, evaporation and several other variables are factored into the equation (Table 1).

Storage Requirements

The effluent storage ponds must be lined with a 20 mil hypalon liner with an underdrain leak detection system.

The effluent storage ponds must be lined with a 20 mil hypalon liner with an underdrain leak detection system. The TWC is now requiring 40 mil liners in most cases along with monitoring wells to insure that leaks are not contaminating ground water supplies. Current regulations also require that aeration be continued while the water is being held in storage. Barton Creek has utilized both fountain and subsurface systems to help increase oxygen levels and prevent stratification of the irrigation water. This is accomplished by the use of diffuser blocks, plastic tubing and small air compressors.

One of our research efforts has been to calculate evapotranspiration rates in an attempt to show that overseeded bermudagrass turf can actually use more water than that allowed in the tables used by the TWC for irrigation purposes.

In 1985 the Texas Water Commission required that the storage pond be sized to accommodate 77 days of storage capacity. The current rule of thumb is 100 days of storage. Table 2 provides an example of how to calculate storage volume. This example incorporates current TWC mandated evapotranspiration indices based on native pasture grass published by Texas Board of Water Engineers in 1960 as Bulletin 6019. Since turfgrass is not addressed in this bulletin, utilization of these indices for wastewater irrigation of golf courses is flawed. The additional construction costs of these required larger capacity ponds may be considerable. One of our research efforts has been to calculate evapotranspiration rates in an attempt to show that overseeded bermudagrass turfs can actually use more water than that allowed in the tables used by the TWC for irrigation purposes in the fall/winter periods.

Figure 1. Maximum Application Rates for Land Disposal of Treated Effluent in Texas (Acre–Feet/Acre/Year)

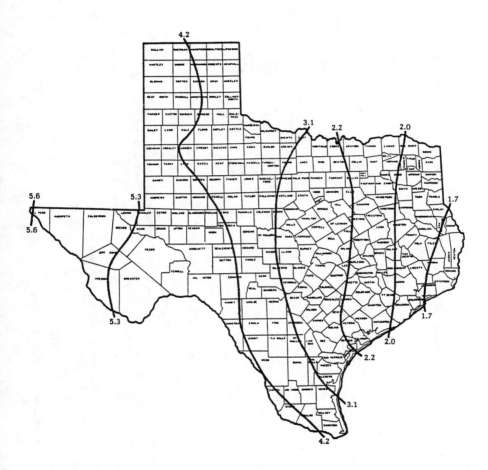

Table 1. Water Balance Example

(Net irrigation effluent need forAustin, Texas for perennial pasture. All units are inches of water per acre of irrigated use.)

Month (1)	Avg. Precip. (2)	Avg. Runoff (3)	Avg. Infiltrated Rainfall[Ri] (4)	Evapotranspiration[b] (5)	Required Leaching[c] (6)	Total Water Needs (5) + (6) (7)	Effluent Needed in Root Zone (7) - (4) (8)	Evaporation from Resvr. Surface[d] (9)	Effluent to be Applied to Land[e] (8)/K (10)	Consumption from Reservoir[f] (9) + (10) (11)
Jan.	1.88	0.34	1.54	0.8	0.00	0.80	0.00	0.087	0.00	0.087
Feb.	3.09	0.56	2.53	1.2	0.00	1.20	0.00	-0.044	0.00	-0.044
Mar.	1.89	0.34	1.55	2.8	0.21	3.01	1.46	0.41	1.72	2.13
Apr.	3.49	0.63	2.86	3.4	0.09	3.49	0.63	0.27	0.74	1.01
May	3.97	0.72	3.25	6.1	0.50	6.60	3.35	0.34	3.94	4.28
June	3.13	0.57	2.56	6.5	0.69	7.19	4.63	0.72	5.45	6.17
July	1.88	0.34	1.54	6.7	0.90	7.60	6.06	1.10	7.13	8.23
Aug.	2.20	0.40	1.80	4.6	0.49	5.09	3.29	1.01	3.87	4.88
Sept.	3.68	0.66	3.02	5.1	0.36	5.46	2.44	0.40	2.87	3.27
Oct.	3.02	0.55	2.47	4.1	0.29	4.39	1.92	0.69	2.26	2.95
Nov.	2.04	0.37	1.67	2.1	0.08	2.18	0.51	0.02	0.60	0.62
Dec.	2.22	0.40	1.82	1.0	0.00	1.00	0.00	0.029	0.00	0.029
	32.49	5.88	26.61	44.4	3.61	48.01	24.29	5.032	28.58	33.612

[a] Runoff should be determined by an acceptable methods such as the Soil Conservation Service Method. Individual storms should be selected from rainfall data for a near average year to obtain an average annual runoff. This annual value is then distributed proportionally to average monthly rainfall to obtain monthly runoff values.

[b] Suggested source of values is from *Bulletin 6019. Consumptive Use of Water by Major Crops in Texas*, Texas Board of Water Engineers.

[c] In low rainfall areas this is the required leaching to avoid salinity build-up in the soil where:

Ce = Electrical conductivity of effluent

$$L = \frac{Ce}{Cl - Ce}(E - Ri)$$

E = Evapotranspiration

Ri = Infiltrated rainfall

Cl = Maximum allowable electrical conductivity of soil solution

Table 2. Sample Calculation of Storage Volume Requirements
(All units are inches of water per acre of irrigated area.)

Month (12)	Effluent Received for Application or Storage (13)	Rainfall Worst Year in past 25 Years (14)	Runoff Worst Year In Past 25 Year (15)	Infiltrated Rainfall (14) - (15) (16)	Available Water (13) + (16) (17)	Net 25 Years Low Lake Evaporation per Acre of Land[c] (18)	Storage (19)	Accumulated Storage[c] (20)
Jan.	2.80	4.88	1.37	3.51	6.31	-0.064	2.76	11.64
Feb.	2.80	4.88	1.37	3.51	6.31	-0.182	2.98	14.62
Mar.	2.80	2.99	0.84	2.15	4.95	0.066	1.72	16.34
Apr.	2.80	5.51	1.54	3.97	6.77	-0.077	2.88	19.22
May	2.80	6.27	1.75	4.52	7.32	-0.072	0.43	19.65
June	2.80	4.95	1.38	3.57	6.37	0.128	-1.59	18.06
July	2.80	2.97	0.83	2.14	4.94	0.348	-3.97	14.09
Aug.	2.80	3.48	0.97	2.51	5.31	0.293	-0.53	13.56
Sept.	2.80	5.81	1.63	4.18	6.98	-0.034	1.32	1.32
Oct.	2.80	4.77	1.33	3.44	6.24	-0.039	1.72	3.04
Nov.	2.80	3.22	0.90	2.32	5.12	-0.026	2.83	5.87
Dec.	2.80	3.51	0.98	2.53	5.33	-0.106	2.91	8.78
	33.60	51.33	14.35	36.98	70.58	0.235		

[a] These values are obtained by totaling Column 11, Table 1 (the yearly amount of consumption of effluent) and dividing by 12. Note that the values in Column 13 could be adjusted to allow for seasonal variation in effluent output.

[b] Individual storms used from the worst year in the last 25 years of data. Total runoff is then distributed proportionally to monthly averages.

[c] Lowest net reservoir evaporation in past 25 years distributed proportionally to monthly average evaporation expressed as inches per irrigated acre.

[d] Storage $= (13) - (18) - \dfrac{(7) - (16)}{k}$. If the term $[(7) - (16)]$ is negative, then the value of the term shall be entered as zero.

Irrigation efficiency, K, is taken as 0.85 for this example.

[e] To allow for the worst condition, the summation of storage was started in September, which gives a maximum storage requirements of 19.65 inches per irrigated acre. This gives an actual reservoir depth of 9.6 feet when reservoir surface is equal to 17% of irrigated land surface.

253

Evapotranspiration Verification

Weather stations and weighing lysimeters were used to verify ET rates based on actual field conditions.

Weather stations and weighing lysimeters were used to verify ET rates based on actual field conditions. The TWC agreed that this is a more accurate method determining rates of disposal, but no changes in the regulations have resulted at this time. We hired Dr. Cornelius Van Bavel, who has worked with evapotranspiration measuring and modification of the Penman Equation since the 1940's. He helped us set up the experiment using weighing lysimeters on both overseeded and non-overseeded bermudagrass in an attempt to establish maximum water usage. Actual water use is measured by weighing turfgrass pots and comparing this data to calculations being made by two different weather stations with different Penman equations. We are approximately one year into this study and hope to release our findings later this year. From this research Toro has just recently changed the way they calculate ET with the Network 8000 system.

Disinfection

A chlorine residual must be present at the point of irrigation.

A chlorine residual must be present at the point of irrigation. This may be attained through the use of chlorine gas or sodium hypochlorite. The water is treated twice when needed, once with gas as the treatment plant discharges into the holding pond and again with liquid as it is pumped into the irrigation system. According to the water commission inspector, 1 ppm constitutes a trace. This sampling is done at the irrigation sprinkler using a Hach kit. Disinfection may also be accomplished by using ozone, ultra-violet light and microwaves.

Salt Problems

One problem occasionally encountered is the buildup of salts in the root zone. This is minimal at Barton Creek thanks to the high quality of the treatment process and the fact that less than 40 percent of the irrigation water is treated effluent. During the latter part of summer we began to see some bicarbonate buildup. Aeration and flushing with large amounts of pH adjusted water is beneficial in reducing these salts levels. Injecting potassium thiosulfate and various microbial agents into the irrigation water to lower the pH, reduce salts, and help kill algae are also under evaluation. Average annual rainfall is 32 inches, which also aids in flushing salts through the soil profile.

Aeration and flushing with large amounts of pH adjusted water is beneficial in reducing these salts levels.

Another problem with the buildup of salts is the potential to cause leaf burn on some types of oak trees, especially Red Oaks and Live Oaks. In an attempt to reduce this damage the irrigation systems were designed to throw water away from sensitive foliage.

Sewage Treatment Plant

The current treatment process meets secondary standards but is about to be upgraded to a tertiary process. Our treatment plant can treat 250,000 gallons per day through an activated sludge complete mix process which generates a high quality water. Water testing shows our quality standards to usually be less than 5 mg/l BOD (Biochemical Oxygen Demand), 5 mg/l TSS (Total Suspended Solids) and 2 mg/L NH_3-N (ammonia nitrogen). Influent is heavily aerated using two large air compressors. This promotes a natural aerobic microbial digestion. From there it moves into a clarifier, where remaining organics are consumed and floating solids are skimmed off the top. The water is chlorinated in a contact chamber and eventually discharged into the holding pond, ready for irrigation.

The current treatment process meets secondary standards but is about to be upgraded to a tertiary process.

Non Point Source Pollution Abatement

As mentioned earlier, non point source pollution abatement programs are required by LCRA when contracting

A Non Point Source Pollution Abatement Plan contains information on water monitoring, IPM, education and golf course maintenance practices.

Bacteria counts in our holding pond are monitored regularly and are usually below 1 colony per liter, which is well under the federal standard for contact recreation of 200 colonies per liter.

Effluent water is provided to us at no charge versus a raw water cost of about 0.90 per thousand gallons.

with this entity for purchase of surface water. The Barton Creek project engineer, Dennis Petras, put together a 400–page Non Point Source Pollution Abatement Plan, which contains information on water monitoring, integrated pest management (IPM), education and golf course maintenance practices. Water testing during storm events is required by the City of Austin to ensure protection of the nearby Edwards Aquifer and Barton Springs, which also serves as a long time swimming area near downtown Austin. Our operation is 7 miles upstream of the springs while there are numerous failing septic tank systems within hundreds of feet of the springs. The City of Austin estimated 30 percent of septic systems will fail over time; however, under our most recent ordinance they exempt all water quality regulations for residential units of 5 acres or more, thereby encouraging the use of individual septic tank systems. Bacteria counts in our holding pond are monitored regularly and are usually below 1 CFU/100 ml (Colony Forming Units/1ml), which is well under the federal standard for contact recreation of 200 CFU/100ml. The only problem detected to–date is the low level phosphates in some storm water runoff. Reducing this concentration by upgrading our treatment plant is underway. Alum will be incorporated to lower the phosphate levels to less than 1 ppm, even though this will remove a nutrient from the water that turfgrass plants could use.

Summary

Effluent water is provided to us at no charge versus a raw water cost of about 0.90 per thousand gallons. Our raw water would normally be drawn from the Colorado River, the drinking water supply for Austin and numerous cities down stream. This is considered an interruptable source for golf course application during emergency water shortage situations and can be discontinued during a severe drought. Effluent water supplies come at a constant rate and are always available. The City of Austin treats its sewage to standards similar to the Barton Creek plant, then discharges this wastewater back into the river.

Even with the numerous regulations and additional maintenance requirements associated with the use of effluent water, it is felt that the benefits in terms of cost savings, water conservation and protection of water quality in the Colorado River far outweigh any of the negatives. Golf course irrigation using effluent can be one of the solutions to Barton Creek water needs in the future.

References

1. Texas Water Commission, Permit to Dispose of Wastes, Permit No. 13206-01, Estates of Barton Creek Utilities, Inc. Approved February 4, 1986.

2. Texas Water Commission, Monthly Operating Report, Permit No. 13206-01, Estates of Barton Creek Utilities, Inc. October 1991 through January 1993.

3. Texas Water Commission, Figure 1 and Table 1 & 2, Rule Book, Chapter 317 Design Criteria for Sewerage Systems, June 1992.

4. Texas Water Commission, Treatment of the Subject, C89-04.

5. Texas Water Development Board, Texas Water Facts, TWDB 91-0166, 1991.

Wastewater Use for Turf: The Non–Technical Economic Side

William S. Rodie
Past President
Arizona Golf Association
Phoenix, Arizona

Wetlands protection, pesticide fate, and underground fuel storage are but a few of the subjects planners, owners, and maintenance personnel must confront.

In an age of environmental awareness, the golf industry faces many challenges. Wetlands protection, pesticide fate, and underground fuel storage are but a few of the subjects planners, owners, and maintenance personnel must confront.

In recent years, there has been increasing pressure, largely in regions where water is perceived to be in short supply, to utilize treated wastewater as an irrigation source for landscaped areas—in particular, turf. Today, the regions with the greatest emphasis on wastewater reuse are Arizona, California, and Florida. While some localized interest exists elsewhere, the most significant lessons are to be learned from these states, and most certainly the experiences from those three states will be watched and will provide lessons for other areas.

Today, the regions with the greatest emphasis on wastewater reuse are Arizona, California, and Florida.

The purpose here is to outline the economic experience of a number of golf facilities that are now using effluent and to quantify in general terms the various costs associated with its use and conversion to its use. As an integral part of the picture, experience in dealing with suppliers is key, and finally, some conclusions on how the golf industry might approach the various aspects of wastewater use to assure a minimal negative economic impact will be explored.

Philosophy

Many feel that wastewater use on turf is the wave of the future, particularly where supplies of ground and surface water are in short supply. Additionally, increasing con-

258

cern for disposal of waste adds credence to finding useful purposes for reuse.

There are two schools of thought on where the burden of paying for waste disposal should fall. The first states that those who create the waste should have the burden of paying for its clean-up. Certainly where industry creates a negative environmental situation, it is, today, asked to bear the cost of clean-up. Likewise then, it would seem reasonable to require homeowners and commercial enterprises that create the waste to pay for its treatment and disposal. To date, that has largely been the case: sewage fees have increased as tougher environmental requirements for disposal have been adopted. It can be stated with certainty that presently the majority of wastewater is disposed of with no secondary use. One can reasonably pose the question, "If it's being thrown away, why not give it away or charge a minimal price to customers who can use it productively?"

The second school says wastewater is a resource, an asset produced by society. Proponents of this outlook state that effluent may ultimately be the only assured source of irrigation for landscape and many agricultural uses - that "primary" water sources may be so short that surface and groundwater resources will be totally dedicated to human needs.

So, we need to ask the question, "Are market forces working to determine the economic fate of wastewater?" The answer is "yes" in a limited number of cases and "no" in most situations.

In California, "special interests" have been successful in creating legislation that now mandates landscape users must convert to wastewater if available and "financially feasible." The feasibility piece has not yet been tested. In Florida, two of the five Water Management Districts now have language in their groundwater use permits stating that the permit will not be renewed if wastewater is available to the permittee - again on a financially feasible basis. In Arizona, wastewater mandates are on a more localized basis. While the Department of Water Resources does impose limits on water use, and does give

Who should pay for wastewater treatment?
- *industry*
- *commercial enterprises*
- *homeowners*
- *recycled water users*

Are market forces working to determine the economic fate of wastewater?

259

*In almost every
instance, the supply of
wastewater is con-
trolled by country– or
municipally–owned
facilities or water
districts either in the
public domain or
privately owned.*

*The golf facility is
placed in a "govern-
mental sandwich"—
mandated to use a
product with one
supplier.*

incentives for effluent use, it does not mandate its use at this time. However, the two largest counties and some cities now mandate wastewater for new turf facilities or, at the least, "renewable" water sources. Groundwater is not a "renewable" source.

In almost every instance, the supply of wastewater is controlled by county– or municipally–owned facilities or water districts either in the public domain or privately owned. In any event, usually a monopoly exists which, as one sage observer so aptly termed it, puts the golf facility in a "governmental sandwich"—mandated to use a product with only one supplier. The only recourse would seem to be effective negotiations with the suppliers or a strongly persuasive appeal to Public Utility Commissions charged with review and approval of rates.

Irrigation Costs for Golf Facilities

In order to evaluate the economics of wastewater, specifically price considerations, it is necessary to review the historical cost of water resources used by golf facilities and overlay the wastewater impact on that picture. In the areas studied, the most common historical supplies came from grandfathered groundwater rights, renewable groundwater rights, surface water (agricultural or other), or water from local municipal or water district purveyors. There are other sources, but the above predominate.

*Historical water
supplies come from
• groundwater rights
• surface water
• local municipal or
  water districts*

Of the sources noted, surface agricultural water is generally the cheapest source - often subsidized and running from 3 cents to 10 cents per 1,000 gallons (see endnotes). Facilities pumping groundwater on-site seem to run 3 cents to 20 cents per 1,000 gallons, largely dependent on electric costs and well depth. It should be noted here that many areas of California have a groundwater extraction tax, which can be as much as 30 cents per 1,000 gallons. Similar legislation has been passed for some areas of Arizona, but is, as yet, not enabled by local governments. Facilities that purchase water from municipalities, water districts, or private water companies may pay anywhere from 40 cents to $4.30 per 1,000 gallons. This latter situation exists in the Monterey Peninsula, and is extreme. A more typical high end would be $2.50 per 1,000 gallons.

To put the above in proper perspective, annual usage should be added. In a hot desert climate, one might expect a 120-acre turfed facility to use perhaps 180 million gallons annually. In cooler areas with more rainfall, usage drops sharply. If our fictional desert facility is fortunate enough to receive 15 cents/1,000-gallon water, its annual cost would be $27,000. However, if the cost is $1.50 for 1,000 gallons, that cost goes to $270,000, an increase of $243,000. Now let's assume 50,000 rounds per year are played, and we see that the cost differential per round is $4.05. This exercise is included here to start putting into perspective some of the impacts wastewater may have on the future of "affordable" golf.

The use of wastewater may have an impact on the future of "affordable" golf.

Facility Front End Costs

Local health departments and state and local environmental departments impose regulations to protect public health and the underlying aquifer. In cases of retrofitting an existing course, signage, backflow protection, and proper markings to distinguish from potable are a must. This is the simplest of conversions, and from reports, runs a minimum of $20,000 if facility labor is used. Others experienced $30,000+ for the basic conversion. In some cases, lake lining may be required to assure the wastewater does not percolate to the aquifer. Best estimates of a plastic lining would seem to exceed $45,000 per surface acre, with total costs dependent on lake configuration and depth.

Simple retrofit or conversion of irrigation systems has run between $20,000 to $30,000 if facility labor is used.

New facilities face similar "extra" costs. In most cases, color–coded pipe is required for wastewater. At this time, due to limited production, a premium of 20 to 30 percent for this material seems to be the norm. Other costs of protection to assure no mixing with potable are similar to those in retrofitting. Additionally, irrigation and drainage designs are required which assure containment to the facility.

261

Problems encountered with system startup include:

- *insufficient pressure*
- *pipe clogging*
- *putting green quality*
- *leaching required*

Perhaps of greater importance than construction requirements are the start–up difficulties that are universal in wastewater use. In no instance of those surveyed was there a smooth start–up. The problems encountered are varied, from insufficient pressure, to system clogging algae, to serious problems on some types of greens which have been all but destroyed. It is impossible to quantify the dollar cost since the problems vary so much, but it can be said that, at present, one must assume some costs due to water quality problems.

In almost all cases, the salts in effluent will require leaching. Again, this is hard to quantify, but a two to five percent increase in water use may be needed.

Of concern is the future quality control of wastewater reuse. If local health and government environmental regulators continue to tighten water quality standards and public health regulations, golf courses using effluent will undoubtably have the attendant costs passed on to them.

Wastewater Costs Considered

Suppliers of effluent are pricing the product with the philosophy that the water is an asset and should be priced near to or at potable rates.

In most areas of California and Arizona, suppliers of effluent are pricing the product with the philosophy that the water is an asset and should be priced near to or at potable rates. There seems to be a general philosophy that 80 percent of potable is "reasonable." There are, however, some very glaring exceptions. In Arizona, the City of Tucson has adopted an 80 percent philosophy, and the City of Phoenix and several other larger suppliers are tending to that same philosophy. That drives a price in Tucson of around $1.10 per 1,000 gallons. Contrast that to a recently opened new facility in a small community near Phoenix that is using local wastewater at a price of 9 cents to 10 cents per 1,000 gallons. In the latter case, the community was anxious to have the golf amenity to help its growth. It had been discarding its effluent, so the sale at a low price attracted the facility and helped with the disposal problem. In California, at least one major golf facility has made a wastewater deal for 5 cents per 1,000 gallons.

One of the best examples of farsighted development is in Orange County, Orlando, Florida. Being inland, disposal of effluent was an obvious dilemma. The problem was solved by building transmission lines to all major landscape areas plus citrus groves and supplying the same with inexpensive or free wastewater. There the attitude was, "These facilities are important to development," and the costs are being borne in sewage fees. Certainly the success of Orange County/Orlando might give rise to other areas' consideration of the value of local development in wastewater use.

One must conclude that local attitudes and conditions will largely dictate not only wastewater prices but also potable pricing where a water company or district is involved in distribution.

One must conclude that local attitudes and conditions will largely dictate not only wastewater prices but also potable pricing where a water company or district is involved in distribution. Certainly, government mandates for wastewater use put the supplier in a stronger negotiation position in making price. While it might appear that golf courses, existing or new, are at the mercy of local water distributors, there are a number of mitigating steps that can be taken.

Action Checklist

1. Whether the facility is to be a retrofit or a new facility, prepare an economic impact evaluation to show the golf facility's economic impact on the area. It is suggested that the National Golf Foundation's *The Economic Impact of Golf Courses on Local, Regional, and National Economies* be used as a guide.

2. Investigate thoroughly the quality of the wastewater product that the supplier is intending to sell you. List all potential difficulties the quality may cause, and try to quantify the cost.

3. Determine what costs may be driven by specific environmental or health mandates.

4. If new transmission lines are to be built, who is responsible for the construction, and how many facilities will be served from the line? If the supplier is to build the line, what type of amortization schedule will be used? Is it reasonable?

5. If wastewater is to mean increased costs for you, can you get local assistance in pleading your case to the water supplier?

6. If planning a new facility, find an area that wants the development and will make concessions. Investigate the circumstances of the area to determine if wastewater disposal is a problem.

7. Would it be worth building the transmission line to get free or very inexpensive water?

8. Obtain quality "at point of delivery" guarantees to assure the supplier has responsibility in any possible action.

All of the points in the Action Checklist are vital in planning and negotiation.

Conclusion

Water planners largely see wastewater as a resource - an asset. In most cases, they do understand that in its various uses, it has problems. They seem to understand conversion and start–up costs plus the ongoing costs of dealing with salts and other problems. While some would sell at the same as potable, most seem to discount to around 80 percent of potable. In California, most golf facilities have been buying water from water districts at prices ranging from 75 cents to $1.50 per 1,000 gallons. The new mandate, with its financial feasibility caveat, has put local water districts in a mode of having to look at the economics of determining if the price of the tertiary treatment plus the amortization of a transmission line would justify conversions. Each case will stand on its own merits, but at this writing, there does not seem to be a rush to build the lines. Most probably, supply of wastewater will be based on studies which indicate that suppliers' cost are such that a price of 80 percent of current potable can be economical to that supplier. In Florida, if the economic feasibility is honored, the conversion may be slower. Reason: most facilities currently are on permitted groundwater, driving low water prices in the

While some would sell at the same as potable, most seem to discount to around 80 percent of potable.

5 cents per 1,000 gallon range, and to build transmission lines and sell at that price would not make economic sense. Arizona mandates, to date, only affect new facilities, so no economic hardship is assumed by pricing the effluent high, since there is no prior supply.

If conversion to treated wastewater is based on economic feasibility, i.e., priced to be reasonably competitive with current prevailing sources, then there should be little alarm as long as "special" costs of using wastewater are not excessive. However, regardless of the type of water available, if prices continue to climb as they have been in California, it will most certainly add to the difficulty of creating reasonably-priced golf so badly needed.

Some very stark facts stand out. One, while most pricing is local and essentially monopolistic, the golf industry must take an active interest in how rates are made and become proactive in that process. Second, it is incumbent on the industry on a local or regional basis to start bonding together to exchange data and help each other overcome the technical and economic difficulties attendant not only to wastewater use, but a variety of other local problems.

Since wastewater is to be a strong irrigating medium for the future, the industry must support ongoing research in low water use grasses and strains that are more tolerant to some of the troubling characteristics of effluent.

The true economic impact of wastewater use will vary greatly at each facility. The most important factor will be how much foresight, investigation, and planning the facility has done. Hopefully, the facility and the supplier will work together to minimize difficulties, and the golf industry, locally, will build a database that will help avoid on-site problems.

The public has a preconceived notion about wastewater—namely that it is "dirty" and "unhealthy." Ironically, many golf courses in the United States historically have been irrigated directly from surface or groundwater sources without any purification. Doubtless, much of this water has been unsuitable for drinking. Yet there have been few, if any, cases of human or animal adverse health

The golf industry
- *must take an active interest in how rates are made and become proactive in that process*
- *must exchange data and help each other overcome the technical and economic difficulties*
- *support research to develop drought or salt tolerant grasses*

Hopefully, the facility and the supplier will work together to minimize difficulties, and the golf industry, locally, will build a database that will help avoid on-site problems.

265

incidents and to our knowledge, there have been no reports of damage to the underlying aquifers from percolation of turf irrigation sources.

The purpose of governmental environmental regulations vis–a–vis wastewater use is to "protect" public health and groundwater purity. The result is regulation on the quality of the product, and its uses tend to assume the worst case scenario before we have enough empirical data to refute the assumptions. Further, the development of the research on the fate of wastewater by–products may well be left to the turf industry to fund. The data on the true impact on public health of wind–blown wastewater spray will take years to accumulate and will undoubtedly be costly.

The development of the research of the fate of wastewater by–products may well be left to the turf industry to fund.

Most certainly in our litigious society of the 1990s, the very existence of regulation will spawn costly legal actions and prompt expensive "overkill" precautions by turf facilities utilizing wastewater. As yet, the insurance industry has seen no need for increased premiums where wastewater is used. One wonders how long that can last.

As yet, the insurance industry has seen no need for increased premiums where wastewater is used.

All this is not to say the turf industry should turn away from using wastewater. In some areas it may truly be the only reliable source of irrigation water, and certainly in some areas it may be a cost-saving factor as well. However, experience to date would indicate it has been more costly. Certainly new facilities mandated to use wastewater are finding it, by and large, costly.

It is incumbent that the industry supports studies and public awareness that can blunt any negative cost factors associated with wastewater use.

At a time when new affordable golf facilities are so badly needed to assure course availability to a broad spectrum of socioeconomic citizenry, it is incumbent on the industry as a whole to support studies and public awareness that can blunt any negative cost factors associated with wastewater use.

Endnotes:

1. Acre Foot (af) - 1 foot of water on 1 acre, or 325,851 gallons.

2. $ Per 1,000 gallons - Cost per 1,000 gallons.

3. $ Per CCF - Price per unit or price per 100 cubic feet of water = 100 cubic feet = 748 gallons.

   Examples:

   a) $1.00 Per 1,000 Gallons = $325.85/Acre Foot

   b) $1.00 Per Unit (ccf) = $435.63/Acre Foot

Appendix Golf Courses Utilizing Effluent Water

Golf Courses Utilizing Effluent Water

Course	Location	
Arrowhead Golf & Country Club	Montgomery	AL
Tierra Del Sol	-	Aruba
Hyatt Regent Coolum	Coolum Beach, Queensland	Australia
Boulders Golf Course	Carefree	AZ
Ocotillo Golf Club	Chandler	AZ
Legend Country Club, The	Glendale	AZ
Wigwam Golf & Country Club	Litchfield Park	AZ
Apache Wells Country Club	Mesa	AZ
Dobson Ranch Men's Golf Association	Mesa	AZ
Leisure World Golf Club	Mesa	AZ
Superstition Springs Golf Club	Mesa	AZ
Foothills Golf Club	Phoenix	AZ
Desert Highlands	Pinnacle Peak	AZ
Desert Mountain	Scottsdale	AZ
Gainey Ranch Golf Club	Scottsdale	AZ
Ken McDonald Municipal Golf Course	Tempe	AZ
Arthur Pack Desert Golf Course	Tucson	AZ
Fred Enke Men's Golf Club	Tucson	AZ
La Paloma Country Club	Tucson	AZ
Silverbell Municipal Golf Course	Tucson	AZ
St. George's Golf Course	St. George's	Bermuda
Chalk Mountain Golf Course	Atascadero	CA
Lakeside Golf Club of Hollywood	Burbank	CA
Spanish Hills Golf & Country Club (1994)	Camarillo	CA
Aviara Golf Course	Carlsbad	CA
La Costa Resort & Spa	Carlsbad	CA
Carmel Valley Ranch Resort	Carmel	CA
El Prado Golf Course	Chino	CA
Western Hills Golf & Country Club	Chino Hills	CA
Eastlake Greens Golf Course	Chula Vista	CA

Course	Location	
Industry Hills Golf Course/Sateuma	City of Industry	CA
Dove Canyon Country Club	Dove Canyon	CA
El Dorado Hills Golf Club	El Dorado Hills	CA
Mile Square Men's Golf Club	Fountain Valley	CA
Pine Mountain Lake Golf & Country Club	Groveland	CA
Skywest Golf Club	Hayward	CA
Meadowlark Golf Course	Huntington Beach	CA
Industry Hills & Sheraton Resort	Industry	CA
Pelican Hills Golf Course	Newport	CA
Castle Oaks Golf & Country Club	Ione	CA
El Niguel Country Club	Laguna Niguel	CA
Lakewood Golf Course	Lakewood	CA
Las Positas Golf Course	Livermore	CA
Recreation Park American Golf Club	Long Beach	CA
Skylinks Municipal Golf Course America	Long Beach	CA
Virginia Country Club	Long Beach	CA
McInnis Golf Center	Marin County	CA
Mission Viejo Golf Club	Mission Viejo	CA
Links at Monarch Beach, The	Monarch Beach	CA
Shoreline Golf Links at Mountain View	Mountain View	CA
Bear Creek Golf & Country Club	Murrieta	CA
Pelican Hill Golf Club	Newport	CA
Oceanside Golf Club	Oceanside	CA
Desert Springs Country Club	Palm Desert	CA
Indian Ridge Country Club	Palm Desert	CA
Portola Country Club	Palm Desert	CA
Palm Springs Golf Course #1	Palm Springs	CA
Palm Springs Golf Course #2	Palm Springs	CA
Brookside Golf Course	Pasadena	CA
Adobe Creek Golf Course	Petaluma	CA
Mt. Woodson Country Club	Ramona	CA
Rancho Murieta Golf Course	Rancho Murieta	CA
Richmond Country Club	Richmond	CA
China Lake Golf Course	Ridgecrest	CA

Course	Location	
San Bernardino Golf Course	San Bernardino	CA
Pacific Golf Club	San Clemente	CA
San Clemente Municipal Golf Course	San Clemente	CA
San Luis Obispo Golf & Country Club	San Luis Obispo	CA
Pendleton Marine Memorial GC	San Luis Rey	CA
San Mateo Golf Club	San Mateo	CA
Santa Barbara Community Golf Course	Santa Barbara	CA
Santa Clara Golf & Tennis Club	Santa Clara	CA
Santa Clara Golf Course Senior Men's Golf Club	Santa Clara	CA
Oakmont Country Club (East)	Santa Rosa	CA
Aliso Creek Golf Course	South Laguna	CA
Bear Valley Springs GC	Tehachapi	CA
Horse Thief Golf & Country Club	Tehachapi	CA
Temecula Creek Golf Club	Temecula	CA
The Sea Ranch Golf Course (in process of switching)	The Sea Ranch	CA
Sherwood Country Club	Thousand Oaks	CA
Tustin Ranch Golf Club	Tustin	CA
Upland Hills Country Club	Upland	CA
Buenaventura Golf Course	Ventura	CA
Olivas Park Golf Course	Ventura	CA
Shadowridge Country Club	Vista	CA
Warner Springs Ranch	Warner Springs	CA
North Ranch Country Club	Westlake Village	CA
Windsor Golf Club, The	Windsor	CA
Ontario National Golf Course	Ontario	Canada
Country Club at Castle Pines	Castle Rock	CO
Amelia Island Plantation/Amelia Links Golf Course	Amelia Island	FL
Beacon Woods Golf Club	Bayonet Point	FL
Casselberry Golf Club	Casselberry	FL
Chi Chi Rodriguez Youth Foundation Golf Course	Clearwater	FL
Feather Sound Country Club	Clearwater	FL

Course	Location	
Cocoa Beach Municipal Golf Course	Cocoa Beach	FL
Daytona Beach Association Golf & Country Club	Daytona Beach	FL
Eagle Pines (will use)	Disney	FL
St. John's County Golf Course	Elkton	FL
Tiger Point Golf & Country Club	Gulf Breeze	FL
City of Jacksonville Beach Golf Course	Jacksonville Beach	FL
Admiral's Cove	Jupiter	FL
Jonathan's Landing Golf Club	Jupiter	FL
Loxahatchee Club	Jupiter	FL
Mayfair Country Club	Lake Mary	FL
Harbour City Municipal Golf Course	Melbourne	FL
Melbourne Golf Course	Melbourne	FL
Spressard Holland Golf Course	Melbourne Beach	FL
Bear's Paw Country Club	Naples	FL
Collier's Reserve	Naples	FL
Flamington Island Club	Naples	FL
Lost Tree Club	North Palm Beach	FL
Eagle Harbor GC (starting in 1994)	Orange Park	FL
Grand Cypress Golf Club	Orlando	FL
Marriott's Orlando World Center Golf Course	Orlando	FL
MetroWest Country Club	Orlando	FL
Wedgefield Golf & Country Club	Orlando	FL
Oceanside Golf & Country Club	Ormond Beach	FL
Stoneybrook Village Golf Club	Pace	FL
Port Malabar Country Club	Palm Bay	FL
Frenchman's Creek Golf Club (1994)	Palm Beach Gardens	FL
Patrick Air Force Base	Patrick AFB	FL
Marsh Landing Country Club	Ponte Vedra Beach	FL
Oak Bridge Club at Sawgrass	Ponte Vedra Beach	FL
TPC at Sawgrass Country Club (Valley Course)	Ponte Vedra Beach	FL
Ballentrae Country Club	Port St. Lucie	FL
Golf Club at Cypress Creek, The	Ruskin	FL

Appendix

Course	Location	
Sanctuary Golf Club	Sanibel Island	FL
Laurel Oak Country Club	Sarasota	FL
Sunrise Golf Club	Sarasota	FL
Lakewood Country Club	St. Petersburg	FL
Mangrove Bay Golf Course	St. Petersburg	FL
Sailfish Point Golf Club	Stuart	FL
Rocky Point Golf Course	Tampa	FL
Wentworth Golf Club	Tarpon Springs	FL
Ironhorse Country Club	West Palm Beach	FL
Winter Springs Golf Club	Winter Springs	FL
Brazell's Creek Golf Course	Reidsville	GA
Settindown Creek Golf Club (late 1993)	Woodstock	GA
Princeville Golf Club	Princeville, Kauai	HI
Hawaii Kai Golf Course	Honolulu	HI
Turtle Bay Hilton Golf Course	Kahuku, Oahu	HI
Kaneohe Clipper Golf Course	Kaneohe	HI
Silversword Golf Course	Kihei, Maui	HI
Makena Golf Course	Kihei, Maui	HI
Mauna Lani Resort	Kohala	HI
Waikoloa Beach Resort	Kohala	HI
Mauna Kea Golf Course	Kohala	HI
Kauai Hyatt Golf Course	Koloa	HI
Kona Country Club	Kona	HI
Experience at Koele Golf Course, The (proposed)	Lanai	HI
Kauai Lagoons	Lihue	HI
Puakea Golf Course	Lihue	HI
Wailua Golf Course	Lihue	HI
Pukalani Country Club	Pukalani	HI
Kuilima Resort	North Shore	HI
Sun Mountain Golf Course	Pahala	HI
White Pines Golf Course	Bensenville	IL
Carbondale Public Golf Center	Carbondale	IL
Fox Run Golf Links	Elk Grove Village	IL
Itasca Country Club	Itasca	IL

Course	Location	
Ivanhoe Club, The	Ivanhoe	IL
Royal Melbourne Golf Club	Long Grove	IL
Wynstone Golf Club	North Barrington	IL
Cardinal Creek Golf Course	Scott AFB	IL
Cantigny Golf Club	Wheaton	IL
Arcadia Valley Golf Club	Ironton	MO
Tournament Players Club at Avenel (scheduled)	Potomac	MD
Cabo Real	Cabo San Lucas, Baja	Mexico
Oak Pointe Country Club	Brighton	MI
Tournament Players Club of Michigan (scheduled)	Dearborn	MI
Excelsior Springs Golf Club	Excelsior Springs	MO
Great Gorge Golf Club	McAfee	NJ
Princeton Meadows Country Club	Plainsboro	NJ
Stanton Ridge Golf & Country Club (in construction)	White House	NJ
Santa Fe Country Club	Santa Fe	NM
Scott Park Golf Course	Silver City	NM
Eagle Valley - East	Carson City	NV
Eagle Valley - West	Carson City	NV
Dayton Valley Country Club	Carson City	NV
Black Mountain Golf & Country Club	Henderson	NV
Legacy, The	Henderson	NV
Royal Kenfield Country Club	Henderson	NV
Desert Rose Golf Club	Las Vegas	NV
Sunrise Golf Club	Las Vegas	NV
Emerald River Golf Course	Laughlin	NV
Heatherwoode Golf Course	Springboro	OH
Shangri-La Golf Course	Afton	OK
Lake Murray State Park	Ardmore	OK
Rock Creek Country Club	Portland	OR
Prineville Meadows Golf & Country Club	Prineville	OR
Summerfield Golf Club	Tigard	OR
Hershey's Mill Golf Club	West Chester	PA

Appendix

Course	Location	
Arthur Hills Course	Hilton Head Island	SC
Long Cove Club (using some)	Hilton Head Island	SC
Moss Creek Plantation	Hilton Head Island	SC
Palmetto Dunes Resort	Hilton Head Island	SC
Palmetto Hall Plantation	Hilton Head Island	SC
Port Royal Plantation	Hilton Head Island	SC
Sea Pines Plantation Golf Club/ Harbor Town Golf Links	Hilton Head Island	SC
Sea Pines Plantation Golf Club/ Ocean Golf Course	Hilton Head Island	SC
Sea Pines Plantation/Seamarsh Golf Course	Hilton Head Island	SC
Shipyard Golf Club	Hilton Head Island	SC
Wexford Golf Club	Hilton Head Island	SC
Oak Point Golf Course	John's Island	SC
Kiawah Island Golf Club	Kiawah Island	SC
Kiawah Island Golf Links/Marsh Point	Kiawah Island	SC
Kiawah Island Golf Links/Ocean Course	Kiawah Island	SC
Kiawah Island Golf Links/Osprey Point	Kiawah Island	SC
Kiawah Island Golf Links/Turtle Point	Kiawah Island	SC
Ellsworth Golf Club	Ellsworth	SD
Legends Club of Tennessee	Franklin	TN
Austin Country Club	Austin	TX
Fair Oaks Ranch Country Club	Boerne	TX
Painted Desert Dunes Golf Course	El Paso	TX
Harlingen Country Club	Harlingen	TX
Kingwood Country Club	Kingwood	TX
Northcliffe Country Club	Cibolo	TX
Buffalo Creek Golf Club	Rockwall	TX
Royal Scot Country Club	New Franken	WI
Christmas Mountain Village	Wisconsin Dells	WI

All golf courses on Monterey Peninsula, CA are scheduled for conversion to effluent upon completion of effluent plant.

Index

Index

C

Distribution facilities, 61-63
Dose-response data, 155, 156
Drainage, 140-141, 242, 243-244
Drinking fountains, 51, 226
Drinking water, 143, 153-157, 183
Drip irrigation systems, 47
Droughts, 19, 24, 26, 54, 111
    avoidance of, 25-26, 27, 32, 33
    resistance to, 25-29
    tolerance of, 26, 28, 32, 33, 45
Dual irrigation system, 50
Dual plumbing, 98
Dyes, 135

E

Echoviruses, 154, 155, 156, 160, 187. *See also* specific
    types
Economic crops, 16
Economics. *See* Costs
Education, 37-38
Efficient use of water, 33
Electrical conductivity, 110, 111
ELISA. *See also* Enzyme-linked immunosorbent assays
Emotional distress, 101
Enteroviruses, 144, 145, 151, 154, 156, 186. *See also* spe-
    cific types
Environmental considerations, 58, 59
Environmental Protection Agency (EPA), 20, 54, 55, 60, 63,
    64-65, 66
    coliform standard and, 151
    guidelines of, 77-86
    health effects and, 143
    microbiological water quality and, 157
    pathogen detection and, 152
    water rights and, 96
Environmental stress, 26. *See also* specific types
Enzyme-linked immunosorbent assays (ELISA), 152
EPA. *See also* Environmental Protection Agency
Escape of plants, 28
Escherichia coli, 187
ESP. *See* Exchangeable sodium percentage
ET. *See* Evapotranspiration
Evaporation, 2, 3, 5, 20, 33
    excess, 35
    reduction in, 134-135
    weather pan, 36
Evapotranspiration, 5, 27, 29, 35, 36, 254
Exchangeable sodium percentage (ESP), 115, 120, 180

H

I

Index

Michigan, 64, 275
Michigan State University Library, 200
Microbiological water quality, 157-158
Microfloculant, 50
Micronutrients, 174. *See also* specific types
Microorganisms, 58, 143, 151, 154, 156, 215. *See also* Pathogens; specific types
    characteristics of, 143-144
    sources of, 144-146
Minnesota, 64
Mission Viejo Country Club, 242-247
Mississippi, 64
Missouri, 64, 164, 275
Molybdenum, 119
Montana, 65, 164
Mulch, 17
Municipal Water Users Association, 130

N

National Pollution Discharge Elimination System (NPDES), 96, 98, 99, 141
Nebraska, 65
Negligence, 99-100
Nevada, 65, 165, 275
New Hampshire, 65
New Jersey, 65, 275
New Mexico, 65, 157, 165, 275
New York, 65
Nickel, 117, 119, 182, 216
Nitrate, 107, 182
Nitrogen, 110, 121, 174, 175, 176, 182-184, 198
Nitrogen loading, 213
Non point source pollution abatement, 255-256
North Carolina, 65, 66, 73-75, 165
North Dakota, 65
Norwalk virus, 145, 151, 156
NPDES. *See* National Pollution Discharge Elimination System
Nucleic acid probes, 152-153
Nutrients, 108, 121, 174, 176, 208-209. *See also* specific types
    availability of, 176-177, 185
    consistent levels of, 198-199
    imbalances of, 30
    macro-, 174
    micro-, 174
    reuse of, 48